Value and Waste in Lean Construction

Non-value-adding activities are otherwise known as 'waste' in the lean construction lexicon. The aim of this collection is to build a common understanding of the role and contribution of value-adding activities in achieving stipulated objectives and continuous improvement in construction projects, and to contrast this with waste. Although the lean approach to the design and the construction of projects has been widely covered as separate issues, there is no book that explicitly provides the link between value and waste in the construction sector: this book bridges the gap. The recognition that waste exists both in design and construction suggests that the realisation of client value depends on how non-value-adding activities are addressed in the architecture, engineering and construction (AEC) sector.

This internationally researched collection showcases high-quality thinking about the link between value and waste. It seeks to create a paradigm shift, which should influence work processes and the way value is conceptualised and operationalised, in both the project management and the business aspects of construction. The readers will gain an understanding of:

- The value-adding paradigm in construction
- How to make value-supporting decisions
- Waste identification and control in practice.

With contributions from Brazil, Denmark, Norway, South Africa, the United Kingdom, and the United States of America, the concepts in this book are globally relevant. The book brings together isolated issues to enhance practice and research. As a reference material for learning, the book emphasises the need to increase value and reduce waste in construction. This is essential reading for all higher-level students of construction management and economics, and all professionals interested in value management.

Fidelis A. Emuze is Associate Professor and Head of the Department of Built Environment, and Head of the Unit for Lean Construction and Sustainability at the Central University of Technology, Free State, South Africa. Construction research in lean construction, health and safety, supply chain management, and sustainability constitutes the main research interest of Dr Emuze, who is a member of the Association of Researchers in Construction Management and the Lean Construction Institute.

Tarcisio A. Saurin is a Professor in the Industrial Engineering Department of the Federal University of Rio Grande do Sul (UFRGS) in Porto Alegre, Brazil. His main research interests are related to lean production and safety management in complex systems. He has worked as a coordinator and/or researcher in funded projects related to those areas in construction, electricity distribution, manufacturing, and healthcare.

Value and Waste in Lean Construction

Edited by
Fidelis A. Emuze and Tarcisio A. Saurin

Routledge
Taylor & Francis Group

LONDON AND NEW YORK

First published 2016
by Routledge
2 Park Square, Milton Park, Abingdon, Oxon OX14 4RN

and by Routledge
605 Third Avenue, New York, NY 10017

First issued in paperback 2021

Routledge is an imprint of the Taylor & Francis Group, an informa business

British Library Cataloguing-in-Publication Data
A catalogue record for this book is available from the British Library

Library of Congress Cataloging in Publication Data
Value and waste in lean construction / edited by Fidelis Emuze and Tarcisio Saurin.
pages cm
Includes bibliographical references and index.
1. Building–Cost effectiveness. 2. Sustainable construction. 3. Construction industry–Waste minimization. 4. Value analysis (Cost control) I. Emuze, Fidelis. II. Saurin, Tarcisio.
TH438.15.V34 2015
624.028'6–dc23
2015022097

Typeset in Baskerville
by Saxon Graphics Ltd, Derby

Publisher's Note
The publisher has gone to great lengths to ensure the quality of this reprint but points out that some imperfections in the original copies may be apparent.

ISBN 13: 978-0-367-35565-4 (pbk)
ISBN 13: 978-1-138-90370-8 (hbk)

Contents

Figures

Tables

Exhibits

Contributors

The editors

Fidelis A. Emuze, PhD, is Associate Professor and Head of the Department of Built Environment, and Head of the Unit for Lean Construction and Sustainability at the Central University of Technology, Free State, South Africa. Construction research in lean construction, health and safety, supply chain management, and sustainability constitutes the main research interest of Dr Emuze, who is a member of the Association of Researchers in Construction Management and the Lean Construction Institute.

Tarcisio A. Saurin, PhD, is a Professor in the Industrial Engineering Department of the Federal University of Rio Grande do Sul (UFRGS) in Porto Alegre, Brazil. His main research interests are related to lean production and safety management in complex systems. He has worked as a coordinator and/or researcher in funded projects related to those areas in construction, electricity distribution, manufacturing, and healthcare.

The contributors

Fabíola Andrade graduated in civil engineering at the University of Fortaleza (UNIFOR – Brazil). Andrade has a specialization in production engineering and an MBA in sustainable construction. Andrade is also a coordinator in Construtora Colmeia (Colmeia Construction Company) and has worked with lean philosophy since 2012.

Lívia Araújo is an undergraduate student in civil engineering at the University of Fortaleza (UNIFOR – Brazil). Araujo is also a trainee at Construtora Colmeia (Colmeia Construction Company).

George Barbosa graduated in civil engineering at the University of Fortaleza (UNIFOR – Brazil). He is an Engineer at Construtora Colmeia (Colmeia Construction Company).

Clarissa Biotto graduated in architecture from Universidade de São Paulo (USP – Brazil). She obtained a Master's degree in construction management and economics from Universidade Federal do Rio Grande do Sul (UFRGS – Brazil). She has authored several national and international papers and is currently an MBA Professor for BIM courses. She is currently the planning director at SIPPRO (Solutions in Planning and Management), a consulting company in construction management, and also a PhD candidate at the University of Huddersfield in the United Kingdom.

Dayana B. Costa, PhD, is Assistant Professor of Construction Management in the Structural and Construction Engineering Department, School of Engineering, Federal University of Bahia – Brazil (UFBA). Dr Costa is the current Chair of Construction Management and Economics Task Group of the Built Environment Technology Brazilian Association (ANTAC).

Brad Hyatt, MS P.E. LEED A.P. BD+C, is Assistant Professor at the Construction Management Program in the Lyles College of Engineering at California State University, Fresno, California. His research interests include leadership in construction, lean construction, healthcare construction management and BIM in construction planning solutions.

Bo Terje Kalsaas, PhD, is a Professor at the Working Life and Innovation Department at the School of Business and Law, and at the Department of Engineering Sciences, Faculty of Engineering and Science, University of Agder, as well as Visiting Professor at the Department of Civil and Transport Engineering, Norwegian University of Science and Technology. His main research and teaching interests are related to supply chain management, lean construction, design management and project management.

Mariana Lima is a PhD candidate in Architecture, Technology and City at the State University of Campinas (Unicamp), and is Assistant Professor of Architecture and Design at the Federal University of Ceará (UFC), Brazil. Mariana developed her PhD thesis focused on how designers and stakeholders should collaborate in order to improve value generation on design.

Søren Lindhard, PhD, is an Assistant Professor at the Department of Mechanical and Manufacturing Engineering at Aalborg University. His research is primarily focused on lean production, planning, scheduling and economics, and he has published articles in journals and presented papers at conferences on these topics.

Sara Costa Maia, Barch, is currently enrolled in the Master of Advanced Studies in Architecture programme at the University of British Columbia, Canada. She is also a Research Assistant in UBC's Transportation

Infrastructure and Public Space Laboratory, and a prior member of the GERCON research group at the Federal University of Ceará, Brazil.

Bruno Mota graduated in civil engineering from the Federal University of Ceará (UFC – Brazil). He is a specialist with an MBA in project management from the Getúlio Vargas Foundation (FGV – Brazil). He has authored several national and international papers and is currently a Professor of MBA courses in management and BIM at INBEC, and is technical director at SIPPRO (Solutions in Planning and Management), a consulting company in lean construction management.

José de Paula Barros Neto, PhD, is Professor and Head of the Technology Center at the Universidade Federal Do Ceara (UFC). Dr Barros Neto is coordinator of GERCON and INOVACON research and consulting groups. His main research focuses are lean construction, production strategy and business strategy.

Alex Opoku, PhD, is currently the Director for the Centre for Sustainability and Resilient Infrastructure & Communities at the London South Bank University, School of Built Environment & Architecture. He holds a PhD in Construction & Project Management from the School of Built Environment, University of Salford in the United Kingdom. Dr Opoku's main research interest is in the area of the sustainable built environment.

André Perroni de Burgos has an MBA in real estate management, and is a Civil Engineer at the Ford Motor Company at Bahia-Brazil.

Rodrigo C. Sanches is a Civil Engineer and a Master's student at the Building Innovation Research Unit (NORIE), Federal University of Rio Grande do Sul (UFRGS).

Bolivar A. Senior, PhD, is Associate Professor at the Department of Construction Management, Colorado State University, Colorado, USA. He has co-authored textbooks in financial management for construction and construction management. Dr. Senior is a member of the American Society of Civil Engineers and the Lean Construction Institute.

John J. Smallwood, PhD, is Professor and Head, Department of Construction Management, and Programme Director, MSc (Built Environment) Programme at the Nelson Mandela Metropolitan University, Port Elizabeth, South Africa. Both his MSc and PhD (Construction Management) addressed health and safety (H&S). He is also a National Research Foundation-rated researcher specialising in construction-related issues such as H&S, ergonomics and health and wellbeing.

Acknowledgements

Many people have been instrumental in the realisation of this book. It builds on the contributions of scholars who pioneer the waste elimination model of lean construction. To mention a few, Lauri Koskela, Glenn Ballard, Gregory Howell, John Rooke, Carlos Formoso, Iris Tommelein, and Luis Fernando Alarcon have laid the groundwork for how waste is perceived and understood in lean construction. The contributions in this book recognise and appreciate the achievements of these pioneers. The support of the secretariat of the International Group of Lean Construction (IGLC) is also appreciated.

The editors of this book have the good fortune to work with colleagues, contractors, consultants and clients who have extended their thinking about waste in construction. Fidelis Emuze acknowledges the contributions of participants in his doctoral research, which inspired continued study of waste in construction. Fidelis Emuze and Tarcisio Saurin are also grateful for the insights about waste in construction gained from discussions with scholars and practitioners in their home countries – South Africa and Brazil, respectively – and abroad.

Anthony Sparg and Renée van der Merwe deserve sincere thanks for their help with English-language editing. The book greatly benefited from the support of Brian Guerin (formerly of Taylor and Francis) and Matthew Turpie of Taylor and Francis. A word of thanks also to all the scholars that freely gave advice and suggestions that benefited the book.

Families have supported and encouraged this work. Fidelis Emuze is indebted to the love and inspiration of his father, Boniface Emuze (1933–2013), his wife, Oluwatoyin Emuze, and their boys, Imole and Irawo. Tarcisio Saurin dedicates this work to his wife, Adriana, and his daughter, Juliana.

Introduction
Goal of lean construction

Why lean in construction?

The interconnection of activities required for the design and construction of buildings and infrastructure involves the interplay between people, technology, situations and decisions. This interplay increases the uniqueness and complexity of a construction product. It also requires the astute coordination of people, materials, tools, plant and equipment to realise the planned progress of work. The coordination is implemented to realise efficiency and enhanced quality in products. However, fatalities, injuries, cost overrun, defects, time overrun, low productivity and many other problems manifest unabated in the construction industry. Improvement of performance is thus imperative for all stakeholders of the industry.

Lean construction has proven to be an alternative for such improvements so as to satisfy clients by creating customer value. The introduction of the concept of lean into the construction sector focuses on the alleviation of design and construction problems by propagating efficiency in decisions and actions. Through its origins in the Toyota Production System (TPS), lean is now applied as an innovative way to manage the design and construction of projects with the use of tools which address project constraints, such as complexities and uncertainties, among others. This application of lean in the construction context is termed lean construction, which has been defined in several ways by notable authors in the discipline. Common accounts of the definitions involve value and waste in construction. The underlying themes of the definitions are oriented in the direction of eliminating waste and creating value. These apparently common themes point to the goal of lean construction.

What is the goal of lean in construction?

Case studies in annual lean construction affiliated conferences, such as the events of the International Group for Lean Construction (IGLC), Lean Construction Institutes (LCI) and the European Group for Lean Construction (EGLC), attest to the perception that lean construction methods can lead to the attainment of both project and business objectives. Lean construction is disseminated as an approach that delivers a construction product by minimising waste to maximise value. In other words, the goal of lean in construction is to apply lean thinking to the design, construction, use and deconstruction or demolition of the built environment. The application of lean thinking by scholars, designers, clients, contractors, suppliers, and the entire construction supply chain targets the attainment of value (customer and supply chain) and removing waste in the construction process. This book therefore highlights the *goal* of lean construction, and its importance for the practice of construction management by focusing on value and waste in lean construction.

Overview of book

This book is designed and compiled to help readers assess their particular circumstances with regard to the adaptation of the waste elimination model in construction. The main idea is that the elimination of waste can be done in various ways that are matched to individual circumstances. The book shows case studies to help readers decide how to adapt an approach to waste elimination in each situation. If a reader is new to the waste elimination model in construction, it would be beneficial to read the chapters in sequence. If the reader is not a novice in the subject area, selective reading can be done within particular chapters. Across ten chapters, the book presents modern-day theory and practice related to value and waste in lean construction. The chapters are positioned under three broad themes: theory of waste in construction; value in construction; and control of waste in construction.

Theory of waste in construction

Part I presents three chapters on waste in lean construction. This part sets out the major ideas of waste elimination in lean construction. Chapter 1 offers insights into the concept and types of wastes in construction. The impact of waste and ways of spotting waste are also highlighted in the chapter. Chapter 2 provides further discourse on variability and making-do. Variability affects the flow of construction work, and could be a source of waste if it manifests at managerial level. However, at the frontline of production, variability may take the form of resilience and thus add value to the process. Although lean construction is focused on eliminating variability in order to eliminate waste, the chapter shows that variability can be positive when it occurs at the frontline of construction operations. Chapter 3 provides insights into the method of measuring workflow while providing a foundation for the implementation of the Last Planner System (LPS) of production control.

Value in construction

A key theme of this book is the relation of value to waste in lean construction. Part II of the book therefore examines this relationship over three chapters. The concept of value as a basis for lean construction and transformation-flow-value theory is presented in Chapter 4. The chapter elucidates attributes of the use of value as the main indicator of waste in lean construction. Chapter 5 takes the value discourse further by providing the relativeness of the concept with decision-making theories. The chapter shows the importance of the role of psychology in the conceptualisation of value in lean construction with easy-to-understand behavioural examples. The synergy that exists when reducing waste to create economic gains through lean and sustainability form the core of the discourse of Chapter 6. The chapter highlights the potentials of shared value in the reconfiguration of the value chain in construction.

Control of waste in construction

Given the notoriety of waste and associated underperformance in construction, Part III of the book presents tools and techniques for its control. Improving planning procedures as an approach for the elimination of waste using LPS as a tool is presented in Chapter 7. The chapter offers insights into how the elements of LPS contribute to the reduction of waste. The use of the kanban system to reduce waste in construction is also presented in Chapter 8. The chapter indicates how to implement the kanban system in construction so as to reduce waste and engender continuous improvement. Chapter 9 is evidence of the use of lean tools in construction. Adapted use of the andon as a visual management tool is shown in the chapter. The andon ensured increased transparency in operations and removed unnecessary interruptions in the case study presented in the chapter. 'Go and see for yourself' as a key concept in lean production is presented in Chapter 10. The technique of genchi genbutsu is crucial to effective determination of waste and its efficient removal in construction.

Taking a pragmatic approach

Practices, tools, methods, and techniques in lean construction are conceptualised and applied in a variety of ways. Throughout this book, the contributions invite the reader to take a pragmatic approach to lean construction application by aligning it to contexts and situations. The alignment should check implementation status at regular intervals and use adaptation required by the idiosyncrasy of each situation to reduce waste in a process that seeks to create value for customers in the construction process. From experience, and from growing empirical research work, the authors in this book have learned that lean construction should be adapted and applied thoughtfully and strategically, in ways that suit the particular context. Not using a mechanical way of applying lean construction could drive the change required for performance improvement in the sector. Prescriptive guides to the application of lean in order to eliminate waste in construction risk creating interventions that are distorted to fit a predetermined format. Such an approach will be contrary to the unique requirements of projects and the customers that commission them. Adaptation of lean in construction thus requires a thoughtful analysis of the situation in which a wide repertoire of tools and techniques, not a one-size-fits-all approach, is used to eliminate waste and create value.

Part I

Theory of waste in construction

1 Wastes in construction

Concepts and types

Fidelis A. Emuze and Tarcisio A. Saurin

This chapter offers insights into wastes in construction. The attributes of wastes are discussed in order to illustrate its impact on value. By means of examples, the chapter explains the concept of waste and how it can have a negative impact on project success. The chapter reviews current literature on waste to argue that activities have to be examined in order to identify, analyse and remove waste, which tends to proliferate if unchecked in the construction process. In other words, project actors should be conversant with the different wastes in construction that occur in the design and production phases of a project, so that strategies can be put in place for the removal of such wastes.

1.1 Background

The construction management literature has alluded to the fact that, in general, performance of projects has remained poor. Time and again, owners and sponsors have asked architects, engineers and contractors to deliver projects within agreed performance parameters of cost, quality, time, health and safety (H&S) and the environment. However, failure to finish projects within these parameters is frequent, and it is usually associated with waste. Just as in the manufacturing industry, waste in construction has negative effects, such as cost overrun, time overrun, low productivity, lack of H&S and lack of competitive advantage (Alwi *et al.*, 2002a, 2002b; Han, 2008; Hwang *et al.*, 2009; Koskenvesa *et al.*, 2010). As a result, waste elimination is a leading principle in lean construction, and a number of lean practices have been used to eliminate waste and enhance value in production environments (Gupta and Jain, 2013). The objective of this introductory chapter is to give an overview of waste in construction, with a discussion of relevant concepts, types of wastes and examples. The discussion thus highlights the need to create awareness of the prevalence of wastes in construction, and the associated impact on project performance.

1.2 Concepts of waste and value

The main goal of lean construction is to eliminate waste from the system by trimming production to make it as value-adding as practically possible. Waste reduction increases production capacity, since the total capacity of a production system is the sum of work and waste in the system (Ohno, 1988). Shingo (1981) defines waste as activities that consume time and resources but fail to add value to the final product. Waste may also be an unwanted physical functionality of the product, as well as the use of more resources than is needed, or an unwanted output. In fact, the concept of waste is inseparable from the concept of value, which means that what is waste for one particular client may not be waste for another client – it all depends on what counts as value for the client. Value is related to outputs that derive from a production system, while waste is related to activities inside a production system, and to unwanted outputs that emerge from the system (Bolviken *et al.*, 2014).

The terms 'value' and 'waste' usually have economic connotations. For instance, value is the price a client pays for a required product or service, and waste is unneeded costs that the client may decline payment for (Han *et al.*, 2007). The costs of waste can take the form of 'direct cost', where production resources are wasted, and 'indirect costs', such as low return on investment, as a result of excessive inventory and low output (Bolviken *et al.*, 2014). It is also worth noting that work that does not add value but is necessary under an operating condition constitutes hidden waste. By contrast, a clear instance of waste is work that does not add value and is not necessary, and where clients' willingness to pay for such work is uncertain. This explanation therefore suggests that value-adding activities should be enhanced; hidden wastes should be made visible and diminished as far as possible; and clear instances of waste should be eliminated in the construction process (Han *et al.*, 2007; Emuze *et al.*, 2014).

1.3 Types of wastes in construction

In a lean production environment, the three broad types of waste are muri, mura and muda.

- Mura is unevenness/non-uniformity, which is variation in work output concerning volume and quality within a production system. Although mura itself is not waste, it leads to muri and muda.
- Muri is overburden, which is described as unreasonable demands on employees or processes, in the form of high workload or non-familiarity with the work to be done.
- Muda is a waste of resources in the form of non-value-adding activity and/or work that creates waste. Such activity contributes to variation in output, and can be controlled through process investigation and removal of root causes (Ohno, 1988).

In construction, Alarcon (1997) used the earlier work of Shingo (1981), which proposes the seven classic wastes, as a starting point. These wastes have informed subsequent classification of wastes in construction (see Table 1.1). It is, however, notable that Sutherland and Bennett (2007) observe that overproduction could be the worst waste among the classic wastes, because it means resources were spent to manufacture unnecessary products, and therefore contributing to the other six classic wastes. In contrast, Koskela *et al.* (2013: 7) contend that 'overproduction is not a dominant waste in construction'. In fact, the classic wastes are context-specific to the manufacturing environment, and, as such, it is crucial to develop a list of wastes that are appropriate in construction (Koskela *et al.*, 2013). This is necessary, as well-known wastes in construction, such as those arising from rework, design errors and omissions and work accidents, are not explicit in the classification of Shingo (1981). Making do, as a waste, has also been proposed in construction. Koskela (2004a) explains that making do occurs when a task has been started before all the preconditions for such an activity have been met. Making do occurs to keep capacity busy, although it usually has detrimental side-effects, such as an increase in work in process, a need for rework and creation of H&S hazards.

Table 1.1 Classification of waste in the production environment

Waste in manufacturing	Waste due to overproduction
	Waste due to stocktaking (taking inventory)
	Waste due to transportation
	Waste due to the system/processing itself
	Waste due to defective products
	Waste due to wait periods
	Waste due to movement
Waste in construction	Work not done
	Rework
	Unnecessary work
	Errors
	Stoppages
	Waste of materials
	Deterioration of materials
	Loss of labour
	Unnecessary movement of materials
	Excessive vigilance
	Extra supervision
	Additional space
	Delays in activities
	Extra processing
	Clarifications
	Abnormal wear and tear of equipment
	Making do

Sources: Alarcon (1997); Koskela (1992; 2004a)

The classification of wastes based on the transformation–flow–value (TFV) theory of production control is another contribution (Bolviken *et al.*, 2014). Transformation-related waste results mainly in material loss; flow-related waste leads to time loss; and value-related waste leads to value loss (Bolviken *et al.*, 2014) (see Table 1.2).

1.4 Identification of wastes in construction projects

Besides a context-specific list of wastes, a project-specific list of wastes can also be produced in order to address the context of a project. Alarcon (1997) proposes a cause–effect matrix that supervisors can use to address wastes on project sites. This matrix can be used by supervisors to capture the views of a project team concerning prevailing wastes, and their causes, on a site. A simple way of doing the work of the matrix is to use a form, such as that depicted in Table 1.3. The form has fields for 'Task/activity', 'Location', 'Project', and 'Date of task', so that background information can be obtained, which may be needed for a better understanding of a particular situation. The form also has a field for 'Observed proceedings',

Table 1.2 A taxonomy of the wastes in the construction production environment

Transformation-related waste	Flow-related waste	Value-related waste
Material waste	Unnecessary movement of people in the work flow	Lack of quality in the main product
Non-optimal use of materials	Unnecessary work in the work flow	Lack of intended use in the main product, which hinders value for the community as a client
Non-optimal use of machinery, energy or labour	Inefficient work in the work flow	Harmful emissions as a by-product
	Waiting in the work flow	Injuries and work-related sickness as a by-product, which hinders value for workers who constitute internal clients in a project
	Unused working space in the product flow	
	Unprocessed materials in the product flow	
	Unnecessary transportation of materials in the product flow	

Source: adapted from Bolviken *et al.* (2014: 820).

where proceedings can be recorded in the form of *genchi genbutsu*, which is one of the five principles of the 'Toyota Way', which enables the problem to be seen firsthand (Marksberry, 2011). In this case, going to the source to find the facts by '*seeing for yourself*' is crucial for corrective decisions that are based on consensus in order to achieve predetermined goals. In other words, the observed proceedings should be based on firsthand knowledge of what has transpired on-site, since an attempt to understand another person's account of the event is always vulnerable to either bias or incorrect interpretation. Observing proceedings assists the site manager or the supervisor to capture the immediate and the remote effects of events. An accurate assessment of actual waste can then be made based on the observed proceedings, and the associated deliberations among project actors. The cause of the waste can also then be discerned from the observations and the brainstorming session among the project actors concerned. The recorded causes of the waste, which must now be removed, should also assist the project team to reach consensus in terms of an intervention strategy.

The hypothetical example provided in Table 1.3 is with reference to the removal of formwork from already-cast columns, as the task that triggered the waste analysis procedure. In Table 1.3, the 'Observed proceedings' field indicates that the columns have voids, which are honeycombs that occur when the mortar has failed to effectively fill the spaces among coarse aggregate particles. While the main problem is concrete casting, the effect of the observed process will cost the contractor more money to fix, because voids are always structural defects that require repairs. Often this cost will end up being passed on to the client. Given that the columns may have to be demolished and recast, the waste that has occurred constitutes rework. One of the causes of the problem could be unworkability of the concrete, and segregation. Excessive bleeding due to loose joints in the formwork can also be a major contributing factor. In correcting the defect, the supervisors in the project would have to undertake the interventions proposed in Table 1.3.

The link between cost and waste in a construction production environment shows the importance of identifying wastes. A case study which used time study and work sampling to observe activities in a precast concrete factory in the USA shows that approximately 47 per cent of the activities conducted in a continuous two-hour period constituted non-value-adding activities (waste) (Nahmens and Mullens, 2011). The activities that made up the 47 per cent included getting tools and materials (25 per cent), rework (15 per cent) and several kinds of waiting periods (7 per cent). The researchers observed that once the activity had started, several issues necessitated flow interruptions. The issues pertained to poor access to materials and tools, rework, poorly defined process flows (sequencing) and supervision matters. Another study, which was conducted in Canada, indicates that many types of wastes can be found on a construction site (Song and Liang, 2011). For instance, Song and Liang (2011) report that a unique sloped-floor design necessitated chipping off of excessive height for about 30 per cent of the columns at the beginning

Table 1.3 An example of a waste analysis form

Waste analysis				
Task/activity: Removal of formwork from columns			*Project:* Low-income housing	
Location: Motherwell, Port Elizabeth, South Africa			*Date of task:* 18-12-2014	
Observed proceedings	*Outcome*	*Waste*	*Causes*	*Interventions*
Striking of the columns reveals huge uneven voids (honeycombs) in the majority of the columns. The honeycombs cannot be corrected with cement grout due to the structural integrity of the building.	Defects; cost overrun	Rework	Unworkable concrete, segregation, congested rebar, inadequate consolidation, and improper concrete-placing practices or workmanship	Ensure that concrete for pouring is workable; review reinforcement details to ensure correct rebar spacing; ensure proper access to formwork; tighten formwork joints; ensure appropriate vibration by workers during placement; train workers to vibrate correctly; ensure that delays when placement is underway are avoided.

of the project. The chipping-off exercise necessitated rework. Rework is a classic example of waste, in which both direct and indirect cost is significant (Love and Li, 2000; Love and Sohal, 2003). Even the indirect cost of rework could be as much as three to six times the cost of rectification (a direct cost) (Love, 2002).

1.5 Summary

Wastes in construction are the nexus of this chapter. The inseparability of the concepts of value and waste was discussed. The different types of wastes, and their characteristics, have been presented. The link between wastes and their causes was established in order to emphasise the tendency of wastes to proliferate in the construction process. Clients in the construction industry like to see work that meets their expectations, and, as such, clients and end users of construction products are interested in value-adding work that ensures that project goals are met by all concerned parties. However, the discourse in the chapter shows that project goals can fail to be achieved if waste is allowed to proliferate unchecked in a project.

Wastes in construction have a relationship with the seven classic wastes. Notwithstanding this relationship, it is crucial for many reasons to develop a comprehensive list of wastes in construction. One reason is the need to action clear improvement measures on-site when a known waste is encountered. Such a list would also promote awareness and consciousness among project actors in terms of wastes in construction. Taxonomy of the wastes in the construction production environment, based on the concept of TFV, is a step in the right direction. In addition, the context-specific nature of construction requires the use of a format for the capturing of prevailing wastes in a project. The wastes that have been cited as examples in this chapter should assist project actors in the identification of wastes in projects. Identification of waste is a crucial first step in developing performance-improvement measures. It has been argued that *genchi genbutsu* should guide the waste-identification process, which is an integral part of waste analysis.

References

Alarcon, L.F. (1997), Tools for the identification and reduction of waste in construction projects. In Alarcon, L. (ed.), *Lean Construction*, Balkema, Rotterdam, pp. 365–377.

Alwi, S., Hampson, K. & Mohamed, S.C. (2002a), Factors influencing contractor performance in Indonesia: A study of non-value adding activities. In: *Proceedings of the International Conference on Advancement in Design, Construction, Construction Management, and Maintenance of Building Structure*, Bali, 2002, pp. 20–34.

Alwi, S., Hampson, K. & Mohamed, S.C. (2002b), Non value-adding activities: A comparative study of Indonesian and Australian construction projects. In: *Proceedings of the 10th Annual Conference of the International Group for Lean Construction*, Gramado, August 2002, pp. 1–12.

Bolviken, T., Rooke, J. & Koskela, L. (2014), The wastes of production in construction. In: *Proceedings of the 22nd Annual Conference of the International Group for Lean Construction*, Oslo, Norway, 23–27 June 2014.

Emuze, F., Smallwood, J. & Han, S. (2014), Factors contributing to non-value adding activities in South African construction. *Journal of Engineering, Design and Technology*, Vol. 12(2), pp. 223–243.

Gupta, S. & Jain, S.K. (2013), A literature review of lean manufacturing. *International Journal of Management Science and Engineering Management*, Vol. 8(4), pp. 241–249.

Han, S. (2008), A hybrid simulation model for understanding and managing non-value adding activities in large-scale design and construction projects. Unpublished PhD thesis, University of Illinois at Urbana-Champaign.

Han, S., Lee, S., Fard, M.G. & Pena-Mora, F. (2007), Modelling and representation of non-value adding activities due to errors and changes in design and construction projects. In: *Proceedings of the 40th Annual Winter Simulation Conference*, Washington, DC, December 2007, pp. 2082–2089.

Hwang, B., Thomas, S.R., Haas, C.T. & Caldas, C.H. (2009), Measuring the impact of rework on construction cost performance. *Journal of Construction Engineering and Management*, Vol. 135(3), pp. 187–198.

Koskela, L. (1992), *Application of the New Production Philosophy to Construction*. Technical Report No. 72, Center for Integrated Facility Engineering (CIFE), Stanford University.

Koskela, L. (2004a), Making do: The eighth category of waste. In: *Proceedings of the 12th Annual Conference of the International Group for Lean Construction*, Copenhagen, Denmark, 3–5 August 2004.

Koskela, L. (2004b), Moving on: beyond lean thinking. *Lean Construction Journal*, Vol. 1, pp. 24–37.

Koskela, L., Bolviken, T. & Rooke, J. (2013), Which are the wastes of construction? In: *Proceedings of the 21st Annual Conference of the International Group for Lean Construction*, Fortaleza, Brazil, 29 July–2 August 2013.

Koskenvesa, A., Koskela, L., Tolonen, T. & Sahlstedt, S. (2010), Waste and labour productivity in production planning: Case Finnish construction industry. In: *Proceedings of the 18th Annual Conference of the International Group for Lean Construction*, Haifa, July 2010, pp. 477–486.

Love, P.E.D. (2002), Auditing the indirect consequences of rework in construction: A case based approach. *Managerial Auditing Journal*, Vol. 17(3), pp. 138–146.

Love, P.E.D. & Li, H. (2000), Quantifying the causes and costs of rework in construction. *Construction Management & Economics*, Vol. 18(4), pp. 479–490.

Love, P.E.D. & Sohal, A.S. (2003), Capturing rework costs in projects. *Managerial Auditing Journal*, Vol. 18(4), pp. 329–339.

Marksberry, P. (2011), The Toyota Way: A quantitative approach. *International Journal of Lean Six Sigma*, Vol. 2(2), pp. 132–150.

Nahmens, I. & Mullens, M.A. (2011), Lean homebuilding: Lessons learned from a precast concrete panelizer. *Journal of Architectural Engineering*, Vol. 17(4), pp. 155–161.

Ohno, T. (1988), *Toyota Production System: Beyond Large-Scale Production*. Productivity Press Inc., Portland, OR.

Shingo, S. (1981), *Study of the Toyota Production Systems from Industrial Engineering Viewpoint*, Japan Management Association, Tokyo.

Song, L. & Liang, D. (2011), Lean construction implementation and its implication on sustainability: A contractors' case study. *Canadian Journal of Civil Engineering*, Vol. 38(5), pp. 350–359.

Sutherland, J. & Bennett, B. (2007), The seven deadly wastes of logistics: Applying Toyota production system principles to create logistics value. Center for Value Chain Research (CVCR) White Paper #0701, Lehigh University.

2 Making do or resilience

Making sense of variability

Tarcisio A. Saurin and Rodrigo C. Sanches

In complex sociotechnical systems (CSSs), such as construction projects, variability of work at the frontline is to some extent unavoidable, and it is often necessary to produce the required outputs. This view of variability is in line with resilience engineering, an emerging paradigm for safety management in CSSs. It complements the traditional focus of lean construction on reduction and coping with variability. However, the boundaries between unnecessary and necessary variability are not clear-cut, and a narrow view of the spectrum of variability types may hinder the identification of waste and innovation. In this chapter, based on insights from both lean construction and resilience engineering, guidelines for making sense of variability of frontline workers are presented. Practical examples illustrate the application of the guidelines, and the consequences of this analysis.

2.1 Background

Lean construction (LC) stresses the reduction of and coping with variability – which is regarded as a source of waste – in internal processes and external suppliers (Koskela, 2000). According to Hopp and Spearman (1996), variability is the quality of non-uniformity of a class of entities, which can be designed into a system (e.g. product variety) or be random (e.g. the time when a machine fails). Story (2011) offers a similar definition, regarding variability as the range of performance measurements, values or outcomes around the average which represents all the possible results of a given process, function or operation. In fact, little variability is a requirement for the use of several lean practices. For example, if variability is high, suppliers are unlikely to replenish stocks just in time, and the downstream production flow may be disrupted. Similarly, the use of fail-safe devices with a shutdown function is not recommended for highly unstable operations (Saurin *et al.*, 2012), as they will often stop working. Nevertheless, the type of variability stressed by LC practices seems to be mostly that of managerial processes related to production and design, rather than the micro-variability of frontline operations. Although the Last Planner System (LPS) of production control, the most well-known LC practice, encourages workers' involvement in the planning of their own work, the focus is on planning more, and 'in greater detail as you get closer to doing the work' (Ballard *et al.*, 2009) in order to reduce variability. However, the level of detail and contents of production plans that fit the macro, and particularly the micro, variability of construction activities have often been taken for granted. An opportunity for LC to place a greater emphasis on the details of work-as-done at the frontline, and to encourage the development of new types of planning, has been opened by the introduction of the concept of 'making do'. According to Koskela (2004), 'making-do as a waste refers to a situation where a task is started without all its standard inputs, or the execution of a task is continued although the availability of at least one standard input has ceased'. Thus, making do implies that work-as-done is different from work-as-imagined in plans. Formoso *et al.* (2011) assert that making do implies a reduction in performance.

In contrast with the lean emphasis, an emerging safety management paradigm called resilience engineering (RE) explicitly values the positive side of variability, particularly that arising from informal working practices, and associated with the performance of frontline workers (Hollnagel *et al.*, 2006). Some exploratory investigations of RE applications to the construction industry have already been made, and opportunities have been identified for the reinterpretation of existing management practices (e.g. Hollnagel, 2014; Saurin *et al.*, 2008). A core assumption of RE is that regardless of the effectiveness of technological and management practices, variability cannot be completely eliminated from CSSs, which are 'intractable' (Hollnagel, 2012). Of course, RE does not argue against the use of practices deemed to reduce variability, as they can support waste reduction. What RE recognises is that such practices are insufficient, and that they need to be reinterpreted to support the necessary and unavoidable portion of variability. Nevertheless, how to identify the desirable threshold of variability (and planning) remains an elusive question, both for RE and LC. This chapter discusses the complementary views of variability taken by LC and RE, emphasising the performance of frontline workers, and implications for the prevention of waste.

2.2 Types of variability emphasised by LC and RE

So far, RE has been mostly a descriptive discipline, characterised by a proliferation of studies reporting stories of resilience, particularly in sectors that involve hazardous technologies, such as aviation and healthcare. Such studies usually describe how the adoption of actions which had not been anticipated by standard operating procedures (SOPs) supported recovery from challenging situations (Righi *et al.*, 2015). Thus, RE studies describe how variability was beneficial to sustain operations, although the need for such variability is often taken for granted as inevitable. Although there is no widely accepted and formalised 'RE classification' of variability types, in this chapter the classification by Hollnagel (2012) is adopted. He classifies the types of variability according to their association with three categories of functions: technological, which are carried out by various types of machinery; human, which are carried out by individuals or groups; and organisational, which are carried out by large groups of people, where the activities are explicitly organised. According to Hollnagel (2012), the default assumption is that technological functions are stable, that human functions vary with high frequency and high amplitude, and that organisational functions vary with low frequency but high amplitude. This means that the variability of human and organisational performance is of most interest to waste prevention. Hollnagel also proposes that the variability of the outputs of functions be classified with regard to time (too early; on time; too late; not at all) and precision (precise; acceptable; imprecise) (Hollnagel, 2012).

As with RE, there is no 'LC classification' of variability types. Koskela (2000) proposes seven categories of requirements that should be in place before starting a task, in order to reduce its variability: design; materials and components; labour; equipment; space; adequate external conditions; and absence of interference from other services. Nevertheless, core LC practices do not seek to understand how, why and when *human performance* varies. While description and measurement of making do offers an opportunity for improving understanding of all types of variability in construction, it has not yet been systematically used for that purpose. New insights into the variability of construction processes could be obtained from the use of cognitive task analysis methods, which are widely used in other sectors for knowledge elicitation, data analysis, and knowledge representation (Crandall *et al.*, 2006). Such methods are still largely under-explored by construction management academics and practitioners (Solis & O'Brien, 2014).

2.3 Making do or resilience

Concerning making do, based on its conceptualisation in this text, its defining characteristic is that it is *waste* that causes a *reduction in performance.* A task being initiated or continued without all its standard inputs is not the major distinguishing characteristic of making do, as resource scarcity is ubiquitous in CSSs (Dekker, 2011). So far, empirical studies (e.g. Formoso *et al.*, 2011) have not been able to quantify the performance reduction associated with making do, although some of its consequences can be described qualitatively, such as a lack of safety. Another difficulty in measuring making do is that it requires an explicit definition of the standard inputs for starting a task. Therefore, it can be questioned whether existing studies are recording making do, or whether they are not perhaps recording another phenomenon.

Concerning resilience, it is defined by Hollnagel (2006) as the 'intrinsic ability of a system to adjust its functioning prior to, during, or following changes and disturbances, so that it can sustain required operations even after a major mishap or in the presence of continuous stress'. While the RE literature does not define precisely what is meant by adjusting functioning, or performance, we propose that it involves one or more of the following: (1) the insufficiency or absence of action rules, which specify in terms of if–then statements how people shall behave (e.g. wearing a seat belt when in a moving car) (Hale and Borys, 2013); (2) improvisation, which is defined by Trotter *et al.* (2013) as the real-time conception and execution of a novel solution to an event that is beyond the boundaries that an organisation has anticipated or prepared for – therefore, improvisation assumes the insufficiency or absence of action rules; and (3) the isolated existence of performance goals and/or process-oriented rules. While performance goals define only what has to be achieved, not how it must be done, process-oriented rules define the process by which the person or organisation should arrive at the way they will operate – e.g. requirements to consult with defined people when an emergency situation arises in order to decide how to handle it (Hale and Borys, 2013).

In fact, resilience is defined as a *functional property* of a CSS, while making do is defined as a system *outcome*. Another difference is that the concept of resilience is neutral, in the sense that it does not specify at which costs required operations are maintained. Waste may or may not occur as a result of resilience. Wears and Vincent (2013) discuss examples of overusing and misusing resilience in healthcare, and the resulting side-effects, such as staff burnout, frustration and resistance to change. Wachs *et al.* (2012) report examples of how resilience may be a way of masking waste in the work of grid electricians.

Both phenomena also share commonalities, such as (1) they are triggered by scarcity of resources; (2) they are emergent, which means they arise from the interactions among several variables, and that they have unique properties that are not found in any of the interacting variables (Cilliers, 1998); and (3) manifestations of both can occur at the individual, team and organisational level. Figure 2.1 summarises the main relationships between the concepts of making do and resilience. According to Figure 2.1, the definition of making do conveys the message that it is intrinsically negative, and that learning from making do equals learning from failure. The figure also indicates that the consequences of working without all standard inputs can only be fully assessed in hindsight, and thus it seems that any measurement of making do cannot be completely conducted in real time.

Considering the aforementioned concepts of making do and resilience, this chapter proposes two categories of performance related to both concepts: (1) unsuccessful resilience, which is equivalent to making do, and corresponds to situations in which performance adjustment implies either waste or safety hazards; and (2) successful resilience, when performance adjustment helps to tackle waste without having any undesired side-effects.

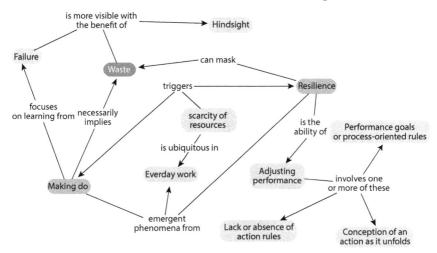

Figure 2.1 Relationships between the concepts of making do and resilience

2.4 An example of unsuccessful resilience or making do

An example of unsuccessful resilience or making do involves the steel structure assembly of a commercial building. The building has nine floors, and 20 workers have been involved in the assembly, which is expected to take four months. The company responsible for assembling the structure is the same company that designed and manufactured the structure. In turn, this company is subcontracted by the main contractor. Data collection involved three visits of one researcher to the construction site, and it included about eight hours of observations of work at the frontline, analysis of SOPs, and interviews with two workers, focusing on the reasons underlying their actions.

Figure 2.2 illustrates the analysed situation, in which a worker is hammering a beam so as to fit it between the columns. This operation is required because the columns and the beam have not been placed in the right position by the crane. However, the worker who is making the adjustments is in an unsafe position as he is using a 5 kg hammer and his feet are on top of the guardrail around the mobile work platform. The hammering could displace the columns even further, which implies rework to reposition the columns after connecting all the pieces.

In this example, making do is characterised by (1) the lack of a standard input for starting the task of connecting beams and columns, namely the fact that there are columns and beams that are not in the right position; and (2) a reduction in performance, represented by the need to use labour to carry out an adjustment under unsafe conditions, and by the rework required to reposition the columns. In turn, resilience is characterised by (1) the lack of any SOP to carry out the adjustment; and (2) a need to

Figure 2.2 An example of unsuccessful resilience or making do

decide how to do the adjustment as the action unfolds – e.g. the worker has to decide, on the spot, how to get access to the beam, and where to hammer. Thus, resilient performance compensates inefficiently for the lack of standard inputs, and it is fully deployed at the individual level, without adequate organisational support. An aggravating condition is that the discussed situation was reported to be normal in all sites of this company, and it is tolerated by management because it has produced the expected outputs. Thus, this is also an example of misusing and overusing resilience.

2.5 An example of successful resilience

A number of examples of successful resilience, related to the pre-assembly of metallic trusses at ground level (Figure 2.3), were identified in the same company mentioned in the previous section. Researchers observed how this activity was carried out on different construction sites, by different groups of workers. Since there was no SOP to guide this activity, execution relied solely on the skills and experiences of workers. Two examples illustrate creative solutions devised by workers: (1) positioning screws on the ground – but without tightening them – reduced work at height, and at the same time allowed some flexibility to move the truss so as to place it in its final

Figure 2.3 Trusses that were pre-assembled at ground level

position; (2) use of knots that would be easy to undo at height, while allowing safe erection of the trusses. These examples show that an SOP does not necessarily has to be interpreted as one of the standard inputs required to start a task in order to avoid making do. Indeed, workers that applied these practices possibly had better performance (i.e. less waste) than workers that did not apply them. Highly experienced and skilled workers are likely to compensate for a lack of SOPs. Of course, after identifying such resilient practices, these should be standardised across all workers, thereby setting a basis for continuous improvement. Identification of waste due to a lack of standardised methods is far from being a new finding in construction, since examples of this can be traced back to the early motion and time studies by Taylor at the beginning of the twentieth century (Taylor, 1982). Last but not least, it is worth mentioning that standardisation does not imply that variability should not be expected. Indeed, in line with lean thinking, workers should be encouraged to be critical of their own practices, and to develop new methods, which could, in turn, be standardised and disseminated as best practices.

2.6 Summary

This chapter presented a comparison between the LC and RE perspectives of variability. The concepts of making do and resilience were useful for the comparison, as they represent manifestations of variability from each perspective. Making do and resilience were found to have commonalities and differences, and for this reason guidelines were offered for the observation and measurement of both phenomena. It was shown that variability is not necessarily a source of waste, since it can also be a source of innovation, and can support a bottom-up approach to the design of better SOPs.

Note

In developing this chapter, the authors have drawn on Saurin and Sanches (2014). The authors gratefully recognise the International Group for Lean Construction in this regard.

References

Ballard, G., Hammond, J. & Nickerson, R. (2009), Production planning control principles. In: *IGLC 17*, Taipei, Taiwan, 15–17 July.

Cilliers, P. (1998), *Complexity and Postmodernism: Understanding Complex Systems*, Routledge, London.

Crandall, B., Klein, G. & Hoffman, R. (2006), *Working Minds: A Practitioner's Guide to Cognitive Task Analysis*, MIT Press, Cambridge, MA.

Dekker, S. (2011), *Drift into Failure: From Hunting Broken Components to Understanding Complex Systems*, Ashgate, London.

Formoso, C.T., Sommer, L., Koskela, L. & Isatto, E. (2011), An exploratory study on the measurement and analysis of making-do in construction sites. In: *IGLC 19*, Lima, Peru, 13–15 July, pp. 236–246.

Hale, A. & Borys, D. (2013), Working to rule, or working safely? Part 1: A state of the art review. *Safety Science*, Vol. 55, pp. 207–221.

Hollnagel E. (2006), Resilience: The challenge of the unstable. In: E. Hollnagel, D. Woods & N. Leveson (eds), *Resilience Engineering: Concepts and Precepts*, Ashgate, Aldershot, pp. 9–17.

Hollnagel, E. (2012), *FRAM: The Functional Resonance Analysis Method. Modelling Complex Socio-technical Systems*, Ashgate, Farnham.

Hollnagel, E. (2014), Resilience engineering and the built environment. *Building Research & Information*, Vol. 42(2), pp. 221–228.

Hollnagel, E., Woods, D. & Leveson, N. (eds) (2006), *Resilience Engineering*, Ashgate, Aldershot.

Hopp, W. & Spearman, M. (1996), *Factory Physics: Foundations of Manufacturing Management*, McGraw-Hill, Boston, MA.

Koskela, L. (2000), *An Exploration Towards a Production Theory and its Application to Construction*, VTT Publications 408, Espoo.

Koskela, L. (2004), Making-do: The eighth category of waste. In: *IGLC 12*, Elsinor, Denmark, .

Righi, A., Saurin, T.A. & Wachs, P. (2015), A systematic literature review of resilience engineering: Research areas and a research agenda proposal. *Reliability Engineering & System Safety*. DOI 10.1016/j.ress.2015.03.007

Saurin, T.A. & Sanches, R.C. (2014) Lean construction and resilience engineering: Complementary perspectives of variability. In: *Proceedings of the 22nd Conference of the International Group for Lean Construction (IGLC)*, Oslo, Norway, 24–27 June, pp. 61–72

Saurin, T.A., Formoso, C.T. & Cambraia, F.B. (2008), An analysis of construction safety best practices from the cognitive systems engineering perspective. *Safety Science*, Vol. 46(8), pp. 1169–1183.

Saurin, T.A., Ribeiro, J.D.L. & Vidor, G. (2012), A framework for assessing poka-yoke devices. *Journal of Manufacturing Systems*, Vol. 31(3), pp. 358–366.

Solis, F. & O'Brien, W. (2014), Critical evaluation of cognitive analysis techniques for construction field management. *Automation in Construction*, Vol. 40, pp. 21–32.

Story, P. (2011), *Dynamic Capacity Management for Healthcare: Advanced Methods and Tools for Optimization*, Taylor and Francis, New York.

Taylor, F.W. (1982), *Princípios gerais da administração científica [General Principles of Scientific Management]*, Atlas, São Paulo.

Trotter, M., Salmon, P. & Lenné, M. (2013), Improvisation: Theory, measures and known influencing factors. *Theoretical Issues in Ergonomics Science*, Vol. 14(5), pp. 475–498.

Wachs, P., Righi, A. & Saurin, T.A. (2012) Identification of non-technical skills from the resilience engineering perspective: A case study of an electricity distributor. *Work*, Vol. 10, pp. 323–333.

Wears, T. & Vincent, C. (2013), Relying on resilience: Too much of a good thing? In: E. Hollnagel, J. Braithwaite & R. Wears (eds), *Resilient Health Care*, Ashgate, Dorchester, pp. 135–144.

3 Measuring workflow and waste in project-based production

Bo Terje Kalsaas

Measuring workflow applied in a strategy for continuous improvement could be an important method for making production of buildings leaner. This chapter summarises a research project, conducted over a period of 3–4 years, where the aim was to find a method to measure workflow with a continuous improvement approach. Two main methods are documented, one based on data gathering by observation, and the other based on individual reports by the workers. Workflow in site production is conceptualised as 'all types of work conducted within available working hours – except obstructions such as downtime, rework and other forms of waste subtracted'. To complement the findings from the research project, the chapter further addresses the method to measure workflow as handover of work between trades, which lays the foundation for the Last Planner System (LPS). The research project delivers extensive empirical material on how time is used on construction sites. The empirical results show a notable amount of waste in several construction projects. This chapter contributes to the understanding of workflow and waste in the production of buildings, and for practical purposes, methods for measuring workflow and observable waste are documented, in order that they can be applied in continuous improvement work at construction sites.

3.1 Background

Two different kinds of construction projects can produce an acceptable profit for the contractor, namely a project that is at a good price but with badly run operations, and a project that is at a bad price but is well run (Kalsaas, 2012) – for instance, if the latter is managed according to the principles of the LPS (Ballard, 2000; Kalsaas & Sacks, 2011). This shows the importance of developing tools for measuring how construction projects are run, in order to establish a basis for making improvements while the project is in progress. This chapter reports on a research project on workflow, and builds on a number of earlier works concerning the definition and measurement of workflow (Bølviken & Kalsaas, 2011; Kalsaas & Bølviken, 2010; Kalsaas, 2010, 2012, 2013; Kalsaas *et al.*, 2014). A number of potential methods to measure workflow are addressed by Bølviken and Kalsaas (2011), and three of these methods are focused on in this chapter. Workflow in construction is discussed by Kalsaas and Bølviken (2010) in relation to Shingo's (1988) concept of flow from manufacturing, which makes a distinction between flow in process and flow in operations. Flow in operations would refer to the flow of work, while flow in process would be the flow of materials. While Shingo (1988) claims that, within manufacturing, flow in process must weigh heavier than flow in operations (workflow), Kalsaas (2013) argues that this lean philosophy may be fruitful in manufacturing, but it is not necessarily so in construction, due to the different characteristics of construction work compared to manufacturing. Some main differences relate to different interdependencies between the trades and/or subcontractors, and the fact that construction is still, to a large extent, craft-based, hence the understanding of workflow is mainly based on Shingo's dimension of operations. Through improved work methods, and reduced variation in work, a higher degree of predictability is expected to be achieved in workflow, which could also improve process flow.

The aim is also to measure workflow within each trade and discipline, not just between the trades, as is the case with handover of work. Furthermore, workflow cannot be understood without at the same time having an understanding of waste, and vice versa (Kalsaas, 2013). Waste in this site production context is understood broadly as downtime and rework, and when related to value, it is in the domain of process value (Kalsaas, 2013). Conceptualisation of workflow is inspired by the concept of overall equipment effectiveness (OEE) from manufacturing (Kalsaas, 2013), and is conceived as made up of three different dimensions, namely smoothness, quality and intensity. Smoothness is expressed by the absence of downtime, quality is expressed by the absence of rework and work intensity is assumed to be constant for measuring periods of approximately one week's duration (Kalsaas, 2013). Workflow in construction is defined by Kalsaas (2013: 5) as 'all types of work conducted within available working hours – except obstructions such as downtime, rework and other forms of waste subtracted'. This understanding of flow is somewhat different from the focus in LPS (Ballard, 2000), where reliable handover of work between trades can be interpreted as being closer to process flow, in Shingo's terms. That LPS-based understanding of workflow is also addressed in this chapter. The main question under scrutiny in this chapter is: 'How can one measure workflow and waste in project-based production?'

The discussion on how to operationalise workflow informs the integration the dimensions of smoothness, quality and work intensity (Kalsaas, 2013). For practical reasons, work intensity is assumed to be constant during the measuring periods. Observable waste, except rework, relates to the smoothness dimension, while rework relates to quality. The most important point in the developed measuring method is to direct the focus towards continuous improvement. For the purposes of analysis, however, data collection also has its merits for enabling benchmarking, if it is applied with great care. When applied to continuous improvement, the idea is to make measurements at different stages in individual projects, and to discuss the findings with the workers, to identify causes of waste, and improvements that can be done. The aim is to understand what goes on at a building site, and to use this information as input to a continuous process of improvement, based on collaboration with and involvement of skilled workers, across disciplines. This should not be confused with Taylorism-inspired time studies, which address how to intensify work.

Waste in construction site production is understood as work and work-related activities which do not add value to the construction under erection, in the form of transformation value, necessary preparations (indirect work, coordination, etc.) for value generation or documentation of technical quality with regard to the owner's specifications and government regulations. Thus, concept of value in this chapter mainly focuses on value in terms of the companies/builders that do the work, not on customer value, although these concepts are not mutually independent. Value for those who do the work is seen as good use of human resources and equipment, and adjustment of the work in ways that maximise efficient use of resources, including minimal waste of working hours. However, this is done not by intensifying the work, but by removing hindrances, such as delays and waiting, which we assume are irritating for the operators.

Figure 3.1 seeks to illustrate conceptual understanding of waste and flow in this chapter. This figure emphasises the causes of waste generation that have a direct impact on workflow. The improvement work seeks to manipulate and influence the mechanisms and drivers that have the capacity to produce waste. The seven well-known wastes of manufacturing (overproduction, time on hand, transportation, processing itself, stock on hand, movement and making defective products) were identified by Taiichi Ohno in his 1978 book *Toyota Seisan Hoshiki*, under the heading 'Complete analysis of waste' (Ohno, 1988: 18). Ohno argues that elimination of these wastes will reduce production costs and increase profit. Ohno sees overproduction as a primary waste, which drives other wastes in a 'vicious circle of waste generating waste' (Ohno, 1988: 55). Koskela *et al.* (2013) discuss these wastes in relation to construction, and they conclude that the main wastes of construction have to be separated from construction itself. Different productions have different logics and mechanisms that we can assume. Bølviken *et al.* (2014) extend Koskela *et al.*'s (2013) argument, where they propose a taxonomy of wastes of production in construction, grouped as the related categories of transformation, flow and value, on which the transformation–flow–value (TFV) theory of production is based. Material losses are related to transformation, time losses to flow and value losses to value.

This chapter addresses flow and time losses. The data which form the basis for this chapter were collected from 2010 to 2012 from observations

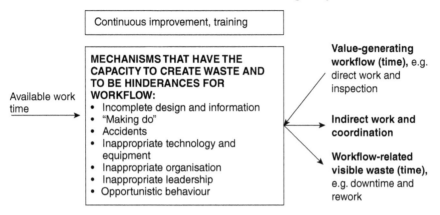

Figure 3.1 A conceptual model for understanding workflow and waste

conducted on building sites in six different projects in Norway, and from activity studies. In one of the measuring methods, based on observation, time studies of activities are applied. The other method of measurement is based on questionnaire self-reports from different work teams. First, the principles of the LPS are explained. The work of earlier time studies is then presented, where direct work, value-adding work and similar items are categorised based on earlier activity studies conducted in the construction industry. A discussion of the data follows, as well as a description of the measuring method used, where the measuring method is evaluated according to how it can contribute to continuous improvement work. Finally, a discussion is presented of the new measuring method currently being tested.

3.2 Last Planner System and flow

The aim of the LPS is to achieve increased control of production in construction, through increased predictability in workflow, and thus reduced waste related to variability. Ballard (2000) describes the LPS system as a workflow control system, which involves coordination of the flow of design, supply and installation through production units. In practice, this method has appeared to work better than traditional methods of planning. Flow, which is a fundamental concept in LPS, is understood as 'the movement of information and materials through a network of production units, each of which processes them, before releasing to next downstream activity' (Kim & Ballard, 2000). In 2014, the Lean Construction Institute (www.leanconstruction.org) defined workflow as 'the movement of information and materials through networks of interdependent specialists'. Expressed differently, we can interpret those definitions as 'handoffs of work between trades'.

The LPS is closely linked to the concept of continuous flow, which is found in the Toyota Production System and in Koskela's (2000) work on the TFV theory (Ballard, 2000), which is considered a theory of production. Koskela's conceptualisation of production consists of three complementary elements, namely transformation through processing at workstations, flow between workstations and the creation of value for the customer or end user.

3.3 Earlier time studies on the proportion of direct work

The purpose of conducting time studies is usually to map what goes on at a building site, with the aim of using this as a starting point for improvement measures. Gouett *et al.* (2011) found that time studies correlate well with improved proportions of direct work when repeated studies are done on a single project. This provides an indication of the usefulness of time studies. A number of time studies have mapped the proportion of direct work on-site. The challenges presented by comparisons are, however, significant, as different studies make use of different categories and definitions of direct work and waste. Heineck (1983) found in a literature study that a number of authors agreed on an average proportion of non-productive time of about 30 per cent in the construction industry, where, for example, Forbes (1977) claimed that the norm would be one-third value-adding work, one-third indirect work and one-third non-productive work. While perhaps not well documented, Mossman (2009) asserts that empirical evidence shows that waste constitutes more than 49.6 per cent of building time, where he defines waste as anything which is not necessary to create value for the customer or end user. Serpell *et al.* (1997), as cited in Josephson and Björkman (2013), found that 47 per cent of building time constituted productive work, while 25 per cent of building time constituted waiting, idle time, travelling, resting and rework. Allmon *et al.* (2000) examined a sample of 72 construction projects in Austin, Texas, from 1973 to 1997, and found that direct work lay in the range of 41–61 per cent, as illustrated in Figure 3.2. Their definition of direct work includes activities such as inspection, clean-up and putting up safety equipment.

In a Swedish study by Strandberg and Josephson (2005), direct value-adding work constituted 17.5 per cent of the work time, while indirect work, material handling and work planning in total amounted to 45.4 per cent. Waiting and unexploited time corresponded to as much as 33.4 per cent of workers' time. Alinaitwe *et al.* (2006) found that 40 per cent of construction work was productive work, where this consists of making the building evolve, preparation of materials, handling materials at the workplace and clean-up/unloading, and that 33 per cent was non-value-adding time (being absent, materials transfer, not working, walking around, waiting and other categories of downtime). In a survey conducted by Skanska Norway (Thune-Holm & Johansen, 2006), the categories 'productive time', 'indirect time', 'change-over time' and 'personal time' were used. In four building projects, the productive time of carpenters was found to be 59.4 per cent, 70.7 per cent, 70.2 per cent and 50.7 per cent, respectively, while for concrete

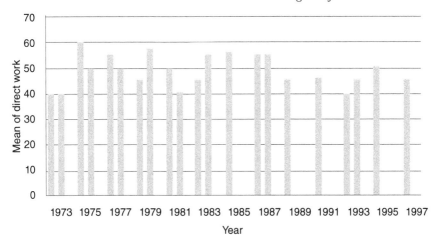

Figure 3.2 Direct work in 72 construction projects in the USA, distributed according to year

workers, productive time was found to be 65.1 per cent and 69.5 per cent, respectively. The method used in this survey is, however, significantly less accurate than the method which forms the foundation of this chapter. Diekmann *et al.* (2004) reviews a series of American studies, of which three were studies of steel erection jobs. The results for the three projects showed, respectively, 32 per cent, 11 per cent and 10 per cent value-adding time, and 60 per cent, 57 per cent and 67 per cent non-value-adding time. Josephson and Björkman (2013) found an average of 13.2 per cent direct work for plumbers, based on a survey of eight projects in Scandinavia. His definition of direct work, however, only included the categories 'assembling' and 'prefabrication on-site', thus making comparisons difficult. A degree of variation in results can be seen in the literature, partly on account of different ways of categorising direct work and waste. In addition, observations show that different trades have different proportions of direct work.

Josephson and Björkman (2013) argue that it is challenging to compare results from work sampling studies over longer periods of time, without considering changes in working conditions, which is an important aspect when we evaluate validity. They claim that the studies are aimed for different uses, and that the papers lack sufficient information on how data were collected; for example, how the observers were trained, and how they interpreted situations. Different definitions and categorisations do, of course, also pose a challenge for validity, when different findings are compared.

3.4 Measurements of workflow in construction

Observation of activities was used as the method of gathering data for measurements of workflow on six different projects within the field of construction. During observation, protocols of activity categories were applied to sample data, which were later aggregated to a total of six categories:

1   direct work: direct work, inspection/control, crane operation and similar;
2   observable waste: direct work → remediation of a mistake, direct work → remediation of a mistake from a different team/trade, waiting/downtime, other personal time;
3   planning, coordination, and health and safety (HS): safety work (HS), planning meetings, coordination and problem-solving on-site, HS meetings;
4   indirect work, logistics: reception of materials, and procedures connected to this, unpacking of materials, delivery of materials to the worksite with a trolley, etc., collection of materials within approximately 12 metres, carrying of waste to the container, displacement between worksites, moving and collecting tools, moving to/from the saw bench in the container, and similar;
5   indirect work, other: rig up and take down, clearing to gain access to the worksite, clean-up after work, and general clearing necessary, and
6   personal time: coffee and lunch break, necessary personal time.

Each recording represents a five-minute period on-site. For further details, see Kalsaas (2013).

The observation template was developed over some time, and, for the purposes of comparing results, some categories have been aggregated in order to achieve consistency in this chapter. The category 'direct work' thus also subsumes the categories 'value-adding demolition' and 'necessary "standby"'. On the same basis, the category 'weather-related rigging' was merged with the category 'rig up and take down'. The categories 'compensatory work' and 'problem solving and other administrative work' were aggregated with the category 'coordination on-site', and the name of the category was changed to 'coordination and problem solving'. The category 'other' was aggregated with the category 'necessary personal time'.

3.4.1 Observation method

Table 3.1 shows the results, distributed according to project, and aggregated for all the projects. The category 'direct work', aggregated on the construction projects, amounted to between 41 per cent and 59.4 per cent, with an average of 49.6 per cent. Included in this amount is 2.1 per cent 'crane operation and similar', and 1.1 per cent 'inspection/control'. Previous time studies show the total of direct work in the construction industry as being in the region of 18–61 per cent, with the average being in the region of 40–45 per cent for all trades. In other words, it would seem

Table 3.1 Overview of aggregated results on different projects

	Secondary school 1: carpenter	Apartment/business/garage: carpenter, electrician	Rehabilitation/facade: bricklayer	Secondary school: concrete/iron	Apartment/garage: carpenter, plumber	Shopping centre: concrete/iron worker	Aggregated results
Number of registration	1,678	2,715	708	3,525	1,267	6,750	
Direct work (%)	59.4	54.1	52.8	57.4	49.6	41.0	49.6
Observable waste (%)	4.6	5.5	6.9	7.4	8.3	15.0	9.9
Planning, condition and HSE (%)	6.0	11.0	3.0	13.3	15.4	15.3	12.7
Indirect work, logistics (%)	8.3	11.9	6.2	9.7	11.2	11.6	10.6
Indirect work, others (%)	2.8	7.0	8.8	2.0	4.8	7.2	5.5
Necessary personal time (%)	18.9	10.5	22.3	10.3	10.7	10.0	11.7
Man hours (hours)	139.8	226.3	59.0	293.8	105.6	562.5	1386.9
Observable waste (hours)	6.4	12.4	4.1	21.7	8.8	84.4	137.3
Calculate workflow (%)	95.4	94.5	93.1	92.6	91.7	85.0	90.1

that the average of the results in this survey is somewhat higher than the proportion of value-adding work reported in the literature. Observable waste constitutes 9.9 per cent, on average, and the variation lies in the range of 4.6–15 per cent, where only one project had more than 8.3 per cent waste. This category is referred to in detail in Table 3.2. Among the individual projects, the shopping centre shows the least amount of direct work, at 41 per cent, and the greatest amount of observable waste, at 15 per cent. In addition, workflow was calculated for each project according to the conceptualised formula of workflow, mentioned earlier, which is calculated as '100 per cent (man hours at employer's disposal – wasted time / man hours at employer's disposal)' (Kalsaas, 2013). The projects measured show large variations in results. This is, however, not surprising: different trades are observed in several of the projects, teams could be performing differently and, according to reports, some teams acted for the study, because of challenges or low performance. In two of the five weeks of observation on the shopping centre, which had the most waste (15 per cent), the researcher was following a team which was performing poorly, with a squad leader who was not showing initiative, and there appeared to be much unproductive time.

In Table 3.2, subcategories of observable waste are shown. Other personal time accounts for 60.3 per cent of observable waste in total, waiting/downtime represents 32 per cent of observable waste, and rework accounts for 7.6 per cent. Rework involves correcting faults, both faults of one's own making and those of others. The researchers find a large proportion of rework in the secondary school 1 project (29.5 per cent) and the apartment/garage project (27.6 per cent) (see Table 3.2).

Table 3.3 presents an overview of the results, distributed across the different trades. Plumbing stands out as the trade which has the smallest amount of direct work, at 13.2 per cent, and a relatively high proportion of observable waste, at 11.7 per cent. The survey by Josephson and Björkman (2013) showed a proportion of only 13.2 per cent direct work for plumbers, which suggests that this trade will perform with a lower proportions of direct work. However, the data sample in this study was small. A high proportion of direct work was found for carpenters, something which is also observed in the survey conducted by Skanska. The highest amount of waste was observed with concrete/iron workers, who had a proportion of 12.4 per cent. Concrete/iron working is the trade with the highest amount of observable waste in this study.

3.4.2 Self-evaluation method

Table 3.4 shows the results of the self-evaluation for the projects in which the method was implemented. Overall, 149 questionnaires were collected from the projects, as follows: apartment/business/garage (119), apartment/garage (21) and rehabilitation/facade (9). Based on the sum of hours of waste, accounted for by the operators, namely 108.9 hours, and the available

Table 3.2 Distribution of observable waste

	Secondary school 1: carpenter	Apartment/business/garage: carpenter, electrician	Rehabilitation/facade: bricklayer	Secondary school: concrete/iron	Apartment/garage: carpenter, plumber	Shopping centre: concrete/iron worker	Aggregated results
Number of registration	1,678	2,715	708	3,525	1,267	6,750	
Observable waste (%)	4.6	5.5	6.9	7.4	8.3	15.0	9.9
Rework (%)	29.5	14.0	0.0	3.5	27.6	4.3	7.6
Waiting/downtime (%)	30.8	7.3	4.1	32.7	7.6	39.5	32.0
Other personal time (%)	39.7	78.7	95.9	63.8	64.8	56.1	60.3

Table 3.3 Overview of aggregated results in different construction trades

	Plumber: apartment/garage	Concrete/iron worker: secondary school 2	Bricklayer: rehabilitation/facade	Electrician: apartment/business/garage	Carpenter: secondary school 1; apartment/business/garage; apartment/garage	Aggregated results
Number of registration	349	10,275	708	276	5,035	
Direct work (%)	13.2	46.6	52.8	50.7	56.5	49.6
Observable waste (%)	11.7	12.4	6.9	4.3	5.6	9.9
Planning, condition and HSE (%)	20.9	14.6	3.0	14.5	9.6	12.7
Indirect work, logistics (%)	19.8	10.9	6.2	12.7	10.0	10.6
Indirect work, others (%)	6.6	5.4	8.8	6.2	5.1	5.5
Necessary personal time (%)	9.5	10.1	22.3	11.6	13.3	11.7

Table 3.4 Self-evaluated waste versus observed waste (%)

	Apartment/ business/garage	Rehabilitation/ facade	Apartment/ garage	Aggregated average
Self-evaluated waste	8.0	9.1	18.4	9.5
Observed waste	5.5	6.9	8.3	6.5

working time for the operators, namely 1,142.8 hours, self-evaluated waste is calculated to be 9.5 per cent in total, based on an eight-hour work day, less 20 minutes to fill out the form. This has to be seen in relation to the amount of average observed waste in the projects in which the self-evaluation method was implemented, which comes to 6.5 per cent. Table 3.4 shows the amount of self-evaluated waste compared to the amount of observed waste for each project.

It should be added that the self-evaluation method was implemented in two studies of a company in the mechanical industry in 2013 (Kalsaas, 2013), where a strong correlation between two methods was found. What is important is that in these two studies it was observed that there was strong motivation in the workforce to participate in the continuous improvement work. The main findings from the observation part of these two studies are reported on later in this chapter. Table 3.5 shows the results with regard to categories of waste from the self-evaluation questionnaires. In the apartment/business/garage project, we see that 'inappropriate equipment' is a significant reason for waste (29.6 per cent). The reason of 'the worksite not being available because of other work' also had a high self-evaluation, at 26.2 per cent. In the apartment/garage project, the reason of 'the worksite had to be cleared before access could be gained' was a dominant reason, at 21.9 per cent, as well as the reason 'performing work today which was not planned prior to the day', and the reason 'spending time correcting faults of one's own or others' (both at 20.2 per cent). In the rehabilitation/facade project, 'missing or inappropriate equipment' was the dominant reason; however, this measurement was based on only nine schemas.

3.4.3 Reliability and validity

Reliability with regard to method, in this case, concerns the question of whether the method affects the data collected. Master's students in their fifth year were responsible for collection of data. Josephson and Björkman (2013) assert that using young, inexperienced observers to collect data is advantageous in this kind of study, since such observers are less biased. However, Jenkins and Orth (2004) argue that observers should be knowledgeable in order to understand what they observe, which could be an issue. The students collecting the data received training beforehand in the method that would be used, and they had good rapport with both the

Table 3.5 Self-evaluation for categories of waste for three building projects (%)

Self-evaluation	Apartment/business/garage	Apartment/garage	Rehabilitation/facade	Aggregated average
(1) Equipment missing or inappropriate.	29.6	10.1	96	28.10
(2) Worksite was not available because of other work.	26.2	25.3	0	24.42
(3) Information was missing or unclear.	12.5	1.7	0	8.80
(4) Worksite had to be cleared before access could be gained.	3.8	21.9	0	8.50
(5) Did you perform work today which was not planned when you started work this morning?	0.9	20.2	0	6.12
(6) Did you spend time today correcting faults or misunderstandings, of your own or others' making?	0.1	20.2	0	5.59
(7) Faulty materials, or inadequate or inappropriate materials.	6.9	0.6	0	4.75
(8) Preceding activity was not completed as promised.	6.8	0.0	0	4.56
(9) Preceding activity was of poor quality, or not completed.	6.4	0.0	0	4.29
(10) Drawings missing, or faults/ deficiencies in drawings.	4.1	0.0	4	2.99
(11) Other causes of delay during work.	2.8	0.0	0	1.88

supervisor and the workers that were observed. Meetings were arranged with the team before and after observation to explain what observation involved, and evaluations were done during working hours, with the students being present to answer questions. However, the researcher cannot in all cases know with certainty which activities are being performed. Some of the observations were done with intervals of several weeks between them, which would give two separate measurements. Weather conditions can, for instance, become a problem. There is, in addition, a great degree of variation in the number of observations within the different trades; accordingly, there will be different degrees of reliability for the results. The rehabilitation and facade project, for instance, had only 708 measurements, and the apartment/garage project also had a relatively low number of measurements at 1,267. The shopping centre had 6,750 measurements, and there is consequently a higher degree of reliability associated with this data, as compared with the other data. However, empirical generalisation is not the main purpose here, but, rather, continuous improvement within each project. The method of observation can affect the findings by influencing the people that are observed, but we have a strong impression that this was not a significant problem in the cases studied, which is confirmed on the construction sites. One could have expected less reliability with the self-evaluation, which is what we found. The choice of method is strongly dependent on the level of motivation of the site crew.

Validity is a question of whether the concepts that have been developed for workflow and waste are useful for capturing the phenomena and processes that they are intended to capture, and that is not empirical generalisation, but rather analytical/conceptual generalisation. One concern could be the question of whether measuring workflow is related to productivity trends. Thomas *et al.* (1984), as cited in Allmon *et al.* (2000), claim that work sampling is a system for indirectly measuring productivity on construction sites by measuring how time is utilised by the workforce. Allmon *et al.* (2000) support this claim, and add that analysing work sampling data collected over a period of time can suggest trends in productivity rates during that period. Josephson and Björkman (2013) conducted a thorough literature review on the possibility that work sampling can be used to predict level of productivity, in the sense that there is a correlation between the amount of direct work invested and the amount of output produced. Most of the literature supports this possibility; however, there is some literature which maintains that work sampling measures utilisation of workers' time, not the productivity level of workers. The main argument is that productivity is dependent on the method and equipment chosen for performing an activity, which means that a new or innovative method or equipment which leads to more value being added to the product over a given period of time does not necessarily change the proportion of direct work (Thomas *et al.*, 1991; Allmon *et al.*, 2000, as cited in Josephson & Björkman, 2013). Direct work is central in our method of measurement, but direct work may also include waste. That is why this study categorises waste as 'observable' in the data sampling, where the implication is that there is waste that is unobservable or hidden in the other categories (Kalsaas, 2013), such as in the 'direct work' and 'indirect work' categories. Making do (Koskela, 2004) is, for example, an issue for those categories of activities. However, the researchers find that the selected categories give a good indication of workflow as it applied to individual projects with the purpose of continuous improvement, in collaboration with the site crew. Furthermore, in the use of the self-evaluation method for this study, it was assumed for practical purposes that the intensity of work remains constant. It is, however, notable that this is not always the case. For example, Seppänen (2009) found in empirical studies evidence of crew slowdowns. Examples are also mentioned in Kalsaas (2013).

3.5 Findings in mechanical construction of oil and gas drilling modules

Operationalisation of the measuring method/instrument should take into account the trades involved. The cited example that was measured (see Kalsaas, 2013) is taken from mechanical work, namely the building of steel constructions which are part of offshore drilling systems, where the floating unit is a ship or a platform. The engineering and construction tasks include drilling decks with heavy drilling equipment distributed across several storeys, as well as mud modules and derricks. Such systems involve extensive pipe systems for pressure regulation, drainage, and hydraulics. The pipe systems are prefabricated in the company's workshop before installation. This kind of production system is subject to major changes during the production phase, often generated by the client via engineering (Kalsaas, 2013). The collaboration across trades that is required in order to ensure access to the different module locations poses a considerable challenge. The measurements reported here were conducted in pipe installations over two one-week periods six weeks apart. Two students observed two operators continuously throughout the working hours every day during the measured periods. Thus the measurements cover two periods of ten days' work – that is, a total of 20 days' work. The data were collected through recording of observed activity every five minutes throughout the day. The activity at each point was generalised to apply for a full five minutes. The categories used in the observation approach largely emerge from Figures 3.3 and 3.4 (Koland & Lande, 2013).

Since there were only small differences between the measured periods, only the aggregated results from both periods are presented here. Figure 3.3 shows the activities during available work hours, distributed between the main categories. This is followed by detailed results for the categories of 'indirect work' and 'waste', which do not appear in the figure. The aggregate results in Figure 3.3 show that more than one in every three hours of

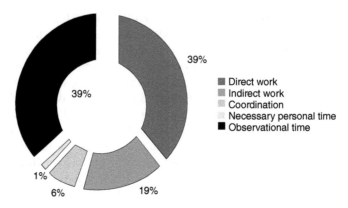

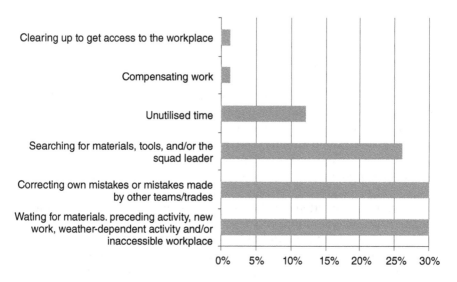

Figure 3.3 Combined results for two one-week measurement periods six weeks apart

Figure 3.4 Distribution of waste categories

performed work is recorded as observable waste. Figure 3.4 distributes the waste according to categories. The study observes that work-related downtime constitutes nearly 45 per cent of the total, and rework constitutes 30 per cent of the total.

Direct work consists of 72 per cent 'direct transformation work' and 27 per cent 'inspection and control', while 'crane operations, etc.' constitutes 1 per cent. Forty-four per cent of the time used for indirect work was spent 'transferring between places of work', and 25 per cent of it was spent 'collecting materials/tools from a distance of more than 5 m'. This shows that there is potential for improving the way the work is organised, as well as minimising observable waste. As part of the measurement regime, a two-part data collection from all of the members of the measured team was conducted. In the first part, the participants were invited to make suggestions for improvement, and various measures were suggested. Contrary to the intentions of this method (see Figure 3.1), however, no systematic improvement measures were implemented on the basis of these suggestions. We underestimated the effort needed to achieve improvement in the reported example.

3.6 Alternative method: handover of work between trades

Following on from the discussion about measuring of operations, according to Shingo's (1988) conceptualisation of flow, we aim to test the definition of workflow found in LPS on reliable handover of work between trades (Kalsaas *et al.*, 2014). Ballard (2000) uses the term 'reliable workflow' in relation to LPS, but one might argue that it is actually process flow which he addresses, if we are to rely on Shingo's (1988) conceptualisation. When one trade has completed its work on a construction object, it may be viewed as a handover of the object to the next trade within the value chain. However, in construction on-site, it is the people that move, not the object. Thus, Ballard's definition of flow is, in this chapter, understood as process flow. An operationalisation to measure process flow on the LPS method is currently being tested. To assess the reliability of handover of work, we developed a questionnaire for the weekly planning meetings (see Table 3.6), where all trades are represented by a foreman and/or a squad leader. In the conceptualisation, we included delays, measured in time units compared to the plan, causes of deviations from the plan, and plan percentage complete (PPC) for handover of work. A condensed version of the questionnaire is presented in Table 3.6. Occasionally there will be more than one trade handing over work to another trade; hence, a separate column is included to indicate number of handovers.

The testing of the method over two weeks shows a total of 44 handovers, where 27 activities took place according to plan, giving a PPC for handovers of 61.4 per cent. One of the findings was that a door was delivered too late in one of the activities, which caused delays in a total of ten work tasks, while the rest of the delayed tasks were caused by re-planning on-site, and a change in the sequence of work tasks. A change in the sequence of work tasks is actually a change in production method. To some extent, this can be expected, as the crews involved acquire new knowledge by experience as they are working.

Table 3.6 An excerpt from the questionnaire: handover of work between trades

From trade/ subcontractor	*To trade/ subcontractor*	*Number of hand overs*	*Handover according to plan?*		*Delays*	*Cause of delay?*	*PPC*
			Yes	No		*Was this notified?*	

3.7 Summary

This chapter shows that the methods for mapping of activity on a building site can be relevant for continuous improvement work in project-based production. This can be achieved through drawing attention to different categories of waste, and through both management and operators becoming more aware of the processes that they are involved in. The method of observation builds on the dimensions of smoothness (a high degree of direct work), quality and intensity. This method gives a fairly accurate picture of what goes on at the building site. It is of value to understand how time is spent at the building site. Such an understanding is of significant interest in production planning. However, there is a need to be cautious in terms of empirical generalisation. The method used and the findings are, in a strict sense, based on research criteria, and they are valid only for individual project sites for the purpose of continuous improvement. This chapter is about analytical generalisation that is noted with case studies. The self-evaluation method of reporting of waste is challenging regarding validity for workflow, and it requires a motivated crew. The self-evaluation method is less work-intensive, which is an advantage compared to the observation method. The self-evaluation method, furthermore, has potential to be integrated into the daily and weekly recording of constraints to workflow for companies applying a standardised piecework wage system.

Calculation of workflow based on the proportion of 'man hours at employer's disposal – wasted time / man hours at employer's disposal' is not the main objective in using the observation method; rather, the main objective is recording of utilisation of time in terms of various different categories as a basis for discussion about constraints and improvement issues. Direct work is central in this method of measurement, but direct work may also include waste. That is why the chapter categorises waste as 'observable' in the data sampling, where the implication is that there is waste that is unobservable or hidden in the other categories. Furthermore, in the use of the self-evaluation method, the chapter assumes, for practical purposes, that the intensity of work remains constant, which is not always the case. One main difference between the methods of observation and self-evaluation and the LPS method of 'handover of work between trades' is that the first two methods also address workflow within the different trades, while the LPS approach only addresses flow between trades. Furthermore, the first two methods are founded on operation, while the LPS method is founded on process, and is based on Shingo's well-known conceptualisation of the different flows in manufacturing. Preliminary testing of the LPS approach is, however, promising, and the contractor company that is hosting the testing is considering expanding its LPS implementation to include the addressed LPS concept of flow for learning and improvement work. A finding from the testing is that the sequence of work tasks may change during the work week, as new knowledge regarding the production method is acquired on-site.

Note

In developing this chapter, the author has drawn on papers published at IGLC conferences (Kalsaas, 2013; Kalsaas *et al.* 2014). The author gratefully recognises the International Group for Lean Construction in this regard.

References

Alinaitwe, H., Mwakali, J.A. & Hansson, B. (2006), Efficiency of craftsmen on building sites: Studies in Uganda. In: *Proceedings of the First International Conference on Advances in Engineering and Technology*, Entebbe, Uganda, pp. 260–267.

Allmon, E., Haas, C.T., Borcherding, J.D. & Goodrum, P.M. (2000), US construction labor productivity trends, 1970–1998. *Journal of Construction Engineering and Management*, Vol. 126(2), pp. 97–104.

Ballard, H.G. (2000), The Last Planner System of production control. Unpublished PhD thesis, University of Birmingham.

Bølviken, T. & Kalsaas, B.T. (2011), Discussion on strategies for measuring work flow in construction. In: *Proceedings of the 19th Annual Conference of the International Group for Lean Construction (IGLC)*, Lima, Peru, 13–15 July.

Bølviken, T., Rooke, J. & Koskela, L. (2014), The wastes of production in construction: A TFV based taxonomy. In: B.T. Kalsaas, L. Koskela & T.A. Saurin (eds), *Proceedings of the 20th Annual Conference of the International Group for Lean Construction (IGLC)*, Akademika forlag, Oslo, Norway, Vol. 2, pp. 811–822.

Diekmann, J.E., Krewedl, M., Balonick, J., Stewart, T. & Won, S. (2004), Application of lean manufacturing principles to construction. Report 191, Construction Industry Institute, University of Texas, Austin, TX.

Forbes, W.S. (1977), *The Rationalisation of House-Building*, Building Research Establishment, Watford.

Gouett, M.C., Haas, C.T., Goodrum, P.M. & Caldas, C.H. (2011), Activity analysis for direct-work rate improvement in construction. *Journal of Construction Engineering and Management*, Vol. 137(12), pp. 1117–1124.

Heineck, L.F. (1983), On the analyses of activity durations on three house building sites. Department of Civil Engineering, University of Leeds.

Jenkins, J.L. & Orth, D.L. (2004), Productivity improvement through work sampling, *Cost Engineering*, Vol. 46(3), pp. 27–32.

Josephson, P.E. & Björkman, L. (2013), Why do work sampling studies in construction? The case of plumbing work in Scandinavia. *Engineering, Construction and Architectural Management*, Vol. 20(6), pp. 589–603.

Kalsaas, B.T. (2010), Work-time waste in construction. In: *Proceedings of the 18th Annual Conference of the International Group for Lean Construction (IGLC)*, Technion-Israel Institute of Technology, Haifa, Israel, 14–16 July.

Kalsaas, B.T. (2012), Further work on measuring workflow in construction site production. In I.D. Tommelein & C.L. Pasquire (eds), *Proceedings of the 20th Annual Conference of the International Group for Lean Construction (IGLC)*, San Diego, Vol. 2, pp. 801–810.

Kalsaas, B.T. (2013), Measuring waste and workflow in construction. In: *Proceedings of the 21st Annual Conference of the International Group for Lean Construction (IGLC)*, Fortaleza, Brazil, 29 July–2 August.

Kalsaas, B.T. & Bølviken, T. (2010), Flow of work in construction: a conceptual discussion. In: *Proceedings of the 18th Annual Conference of the International Group for Lean Construction (IGLC)*, Technion-Israel Institute of Technology, Haifa, Israel, 14–16 July.

Kalsaas, B.T. & Sacks, R. (2011), Conceptualization of interdependency and coordination between construction tasks. In: *Proceedings of the 19th Annual Conference of the International Group for Lean Construction (IGLC)*, University of Salford, Manchester, pp. 33–44.

Kalsaas, B.T., Gundersen, M. & Berge, T.O. (2014), To measure workflow and waste: A concept for continuous improvement. In: B.T. Kalsaas, L. Koskela & T.A. Saurin (eds), *Proceedings of the 20th Annual Conference of the International Group for Lean Construction (IGLC)*, Akademika forlag, Oslo, Norway, Vol. 2, pp. 835–846.

Kim, Y.W. & Ballard, G. (2000), Is the earned-value method an enemy of work flow? In: *Proceedings of the 6th Annual Conference of the International Group for Lean Construction (IGLC)*, Brighton.

Koland, S. & Lande, T. (2013), Lean orientert analyse av arbeidsflyt og waste i prosjektstyrt mekanisk industri med levering til offshore olje- og gassmarkedet. Case AS Nymo [Lean-oriented analysis workflow and waste in a project-based mechanical industry with delivery to offshore oil and gas market. Case AS Nymo]. Unpublished Master's thesis, Department of Working Life and Innovation, University of Agder, Grimstad.

Koskela, L. (2000), *An Exploration Towards a Production Theory and its Application to Construction*, VTT Publications, Helsinki.

Koskela, L.J. (2004), Making do: The eighth category of waste. In: *Proceedings of the 12th Annual Conference of the International Group for Lean Construction (IGLC)*, Copenhagen, Denmark, 3–5 August.

Koskela, L., Bølviken, T. & Rooke, J. (2013), Which are the wastes of construction? In: *Proceedings of the 21st Annual Conference of the International Group for Lean Construction (IGLC)*, Fortaleza, Brazil, 29 July–2 August.

Mossman, A. (2009), Who is making money out of waste? Unpublished research proposal, Loughborough University.

Ohno, T. (1988), *Toyota Production System: Beyond Large-Scale Production*, Productivity Press, Portland, OR.

Seppänen, O. (2009), Empirical research on the success of production control in building construction projects. PhD dissertation, Helsinki University of Technology.

Shingo, S. (1988), *Non-stock Production: The Shingo System for Continuous Improvement*, Productivity Press, New York.

Strandberg, J. & Josephson, P.E. (2005), What do construction workers do? Direct observations in housing projects. In: *Proceedings of the 11th Joint CIB International Symposium: Combining Forces: Advancing Facilities Management and Construction through Innovation*, Helsinki, 13–16 June.

Thomas, H.R. (1991), Labor productivity and work sampling: The bottom line. *Journal of Construction Engineering and Management*, Vol. 117(3), pp. 423–444.

Thomas, H.R., Guevara, J.M. & Gustenhoven, C.T. (1984), Improving productivity estimates by work sampling. *Journal of Construction Engineering and Management*, Vol. 110(2), pp. 178–188.

Thune-Holm, E.C. & Johansen, K. (2006), Produktivitetsmålinger i Skanska [Productivity measurements in Skanska]. Internal Skanska Report, Oslo.

Part II
Value in construction

4 A systemic approach to the concept of value in lean construction

Sara Costa Maia, Mariana Lima and
José de Paula Barros Neto

The concept of value is fundamental in lean construction and transformation–flow–value (TFV) theory. This study aims to develop a solid and coherent theoretical background for the concept by providing an extensive literature review, not only of the field of lean construction, but also following a systemic approach in several areas of knowledge. We discuss the complex implications that the study of value presents, including with regard to sustainability and ethical responsibility. This study contributes towards a better understanding of the potential and the pitfalls of the use of value as a main indicator of waste in lean construction.

4.1 Background

Identifying and managing people's evaluative process – and, ultimately, value – is a matter that engages the interest of several areas of knowledge, from economics to experimental psychology (Fischhoff, 1991). It accompanies the evolution of social movements and scientific paradigms, reflecting one of the most prominent post-Second World War concerns, namely the individual (Davis 2013). In construction, the importance of the concept of value gained relevance when the concept of lean thinking (LT) was integrated into the new discipline of lean construction. Client value identification is a key stage in the philosophy of LT. Without value identification it would not be possible to distinguish waste, or, consequently, to enable evolution in production processes.

In lean construction, the concepts of value and value generation are presented by the TFV theory of production. This theory integrates the concepts of production associated with transformation, flow and value. According to Koskela (2000), in the first concept – transformation – production is seen as the transformation of inputs into finished products, and overall process improvement is obtained by improving the sub-processes individually. In the second concept, flow, activities to connect the sub-processes – the flow activities – must be considered as well, e.g. moving, waiting and inspection. However, except in very specific cases, such activities do not add value to the product, which suggests as the main starting point for production improvement the reduction of non-value-adding activities, thus eliminating waste. In the third concept – value – production incorporates the client. In its essence, production value can only be set contingent on the client. The input for production processes is derived from clients' dependent information and based on their needs and requirements. Thus, waste can only be distinguished in a given set of client evaluative parameters.

Considering TFV theory applied to design, the concept of value becomes even more important. The interaction with clients expected for this phase of the construction process makes the concept of value more relevant to design than the concepts of transformation and flow (Koskela 2000). Also, it is in the design phase when generation of value for a building can be the most efficient. Currently, however, little emphasis has been given to a consistent and internally coherent understanding of value in lean construction. Erikshammar *et al.* (2010) discuss the ambiguities found in the discourse on lean construction pertaining to the concept of value. In fact, their discussion reproduces the same divergences pointed out in the fields of marketing and economics by Woodruff (1997).

After conducting a comprehensive literature review, in LT and in several other areas of knowledge, it became apparent that these conflicts can be explained by indiscriminate assimilation of propositions, as well as artificial categorisations that the concept of value has acquired over time. Often, these propositions and categorisations regarding the concept of value are taken out of context. The analytical methods adopted often fail to support the purposes of lean construction; on the contrary, they raise discussions and contradictions that are fundamentally irrelevant, and impair a general understanding of a concept that otherwise could find alignment in contemporary thinking throughout a number of disciplines.

As classic examples of the type of categorisations mentioned above, Aristotle's and Adam Smith's propositions can be cited. Aristotle sub-categorised value into economic, political, moral, aesthetic, social, legal and religious value. Adam Smith (and later Karl Marx), in turn, suggested the famous distinction between use value and exchange value. In lean construction, Thyssen *et al.* (2010) also suggest a distinction between the notions of value and values, while Rooke *et al.* (2010) propose a clearer difference between sociological value and economic value. Now when defining the concept of value, lean construction and value management authors resort to Adam Smith and the classical school arguments, although already in the nineteenth century the Classical School of economics had suffered severe criticism and its theories of value were taken as inadequate (Feijó, 2001).

Menger (2007),[1] by the end of the nineteenth century, had already rejected the main distinction between use value and exchange value, and assumed that these concepts are subordinated to the concept of value itself. The authors of this chapter argue that the distinction between values and value also shows no practical use for this work, particularly when it aims to select objective, predictable and measurable aspects of value over less understood aspects of the concept. Note that what might be called 'values' or 'moral values' or 'social values' are the main regulators of habits, and therefore of human activities in general, which the built environment must necessarily respond to. It can be observed that the formulation of moral judgements fulfils important regulatory functions of the individual's own behaviour, and of the behaviour of others, through social interaction, and competes strongly with the way a given attribute is evaluated by an individual. Rooke *et al.* (2010) also assert that economic value is one kind of sociological value, which starts to suggest a concept of value less fragmented than what is often discussed. One of the motivations of this study is to refute the use of hierarchical schemes that selects certain artificial categories of value, such as those of economic reference, over artificial categories of a non-measurable nature. It is herein argued that such an approach may detract from lean construction's objective of identifying waste through the way clients or users weight and evaluate the attributes of a building, regardless of its objective performance.[2]

Ultimately, this chapter intends to propose a comprehensive definition of the concept of value from a systemic perspective which is pertinent to the current discussions being conducted in lean construction. It starts by exploring a cohesive set of aspects that describes the concept's behaviour in order to provide further understanding of the proposed definition. These aspects are discussed in the section where a basic understanding of value is introduced. Because an understanding of value is incomplete without analysing how value expresses itself in the world, further discussion is carried out concerning choices and preferences. After this, it became possible to propose a comprehensive definition for value, in which important constructs, such as value generation, can be defined in a coherent and internally consistent manner inside lean construction. These thoughts are summarized in Section 4.5. Based on the features of value discussed throughout the text, the chapter challenges assumptions concerning the relationship of value with welfare, ethical responsibility and sustainability in lean construction. Discussions on these topics are initiated in Section 4.6. Finally, the summary poses the question: should value be really taken as the ultimate measure for waste?

The chapter argues that in order to achieve comprehensive and constructive understanding of value, a systemic approach that acknowledges the complexity of the concept's nature and dynamics is needed. Therefore this chapter seeks a systemic understanding of value by presenting a critical review of the concept of value throughout a number of disciplines, making use of a research strategy grounded in the concept of complexity.

4.2 Problem approach and methodology

This study has a theoretical approach which is based on a systematic and transdisciplinary perspective of value. It reviews complementary studies from the fields of economics, philosophy, psychology, neurosciences, and semiotics, among others, from a complex[3] perspective. Convergent aspects of value are interrelated and synthesised, in terms of the TFV proposition of the concept. In this way, this chapter aims to contribute to the development of a more consistent background for the concept of value in lean construction, thereby enabling further questions to arise. For this purpose, the chapter uses as a point of departure the hypothesis that the concept of value, as well as the design process and its relation with client values, is complex, and requires a systemic approach in order to overcome the unproductive ambiguities often found in lean construction literature.

The chapter refers to a systemic approach based on the framework proposed by Lima *et al.* (2011) in their paper 'Epistemological basis of design process: Positivism versus complexity'. This approach is based on the assumptions of complexity, instability and intersubjectivity, related to what Vasconcellos (2002) calls the 'emerging new-paradigmatic science', or just 'new-paradigmatic science', referring to a (twentieth-century) postmodern science. Lima *et al.* (2011) summarise these concepts as follows:

> Complexity: contextual, search for recursive causal relations.
> Instability: the world in the process of becoming. The phenomena are unpredictable, irreversible and uncontrollable.
> Intersubjectivity: impossibility of objective knowledge of the world, multiple views of reality.

The concept of complexity developed by Morin (2007) and suggested by Lima *et al.* (2011) is one of the main drivers of this chapter. Together, these concepts were used as reference filters in an attempt to bring together a wide range of fields in the definition of the concept of value according to the purposes of lean construction. In practical terms, this methodological approach implies a few things. The first of them is that an extensive and transdisciplinary literature review has been conducted based on the problem, i.e. understanding value as supported by the TFV theory. The chapter does not address the problem inside disciplinary silos, but rather it is searching for conceptual and empirical explorations that build upon a shared view of the concept, consonant with lean construction's needs. The chapter compiles a comprehensive understanding of value's nature and dynamics with the exploration. Additionally, readers should note that the description of value curated in this chapter is imperfect, and that subjective reasoning is employed to filter and process the literature surveyed.

Finally, and most importantly, the chapter to some extent recognises the complexity in the concept of value, as well as the instability and intersubjectivity of the world. Therefore, the conducted literature review focused on studies that shared a view of value that is inherently complex. In the following section, the findings from an extensive literature survey are presented in the form of a description of value as a phenomenon. This description is broken down into smaller subsections for didactic purposes, exploring specific aspects of value. It approaches only a fundamental understanding of the concept, laying the basis for the next discussions on practical manifestations. A final definition is suggested after fundamental and instrumental characteristics of value have been discussed.

4.3 A critical review of the concept of value

4.3.1 The nature of value

There is a strong relation between value and the client's judgements, needs, wishes and satisfaction, which are recurrent in literature. Koskela (2000), for instance, states that value refers to the fulfilment of customer requirements. However, it must be noted that despite their close relation, these concepts are not equivalent. As this chapter moves forward towards a definition of value, it is expected that some of the distinctions between terms will become evident.

For a start, the chapter explores value regarding its nature. In a general appreciation of the concept, Garcia (2003: 105) states that 'value is a relation established between subject and object'. This statement raises important points regarding the nature of the concept of value. First of all, value is a relation. According to Wolfgang Köhler (1938), value must not be equivalent to useful or convenient, as nothing is useful or convenient in and of itself. An object's utility, or convenience, points to something beyond, something that is mainly acquired. Thus, value may constitute precisely this referential relation. Menger (2007)[4] also argues that value is not inherent in goods, but that it is, in fact, the importance that certain goods acquire for users. Thus, value is 'the importance that we first attribute to the satisfaction of our needs and that we transfer to economic goods' (Feijó, 2001: 389). Following the same logic, Köhler (1938) suggests, in a phenomenological analogy, that value has a vectorial nature. The vector of value is experienced as issuing from a definite part of the field, the subject, and is directed towards the referenced object. So, through such vectors, the subject either accepts or rejects the object as valuable. Garcia (2003: 97) emphasises the importance of adopting the concept of field in this relation. In this context, value is no longer presented as something isolated, as resulting from a polarised relationship between subject and object, but as something that is inserted in a sensitive field of action. This means that the evaluation of an object is also relative to the relation that the subject establishes with other objects, with other subjects (intersubjectivity), and with society (social macro-subject), including framing matters and external conditions. Thus, all these elements constitute a system. From this point of view, other elements within the system – or the context – also influence the subject–object relation. Slight changes in this context, and changes are constantly occurring, make the evaluative activity dynamic, and even unstable. The subject is not a constant entity either, as it transforms subtly and continually, informed by experience. These aspects of value will be discussed further in the following sub-sections.

4.3.2 The instability

It is critical to consider the volatile nature of value. This nature requires the correct identification of the system, of the temporal and spatially located environment, as an important component not only for identifying value in an architectural object, but also for the conceptualisation of value itself. It is necessary in the conceptualisation of value that the inherent dynamic characteristics of value be understood, and not denied or presented as if they are an impediment to proper definition of the term.

Rittel (1966: 29), on the consequences of decision theory, argues that 'the systems of values can no longer be regarded as established for long periods of time. What is wishful depends on what may become possible, and what should become possible depends on what is wished'. Thus, to be successful, long-term strategies should not be offered through inflexible decision models. When value is considered as a vector in a field of action whose entities are always changing, it is assumed that even if it were possible to accurately quantify and qualify the value that an individual assigns to an particular object at a given moment, it cannot be ensured that such properties will remain identical for any other moment later, or would have been identical earlier. Thus, in the case of an attribute of a building, or the building as a whole, one can say that the value a particular user assigns to the building is in constant change: from the moment a client analyses a product on offer, to the moment they make the purchase, to their first use experience, and then the second, and then the third, to the development of ties of affection and ownership, to their immediate needs, to their twentieth use experience. Each new context implies a new valence of value.

This quality of spontaneous evaluation processes has recently been acknowledged by product designers. Traditionally, product evaluation practices would focus on users' early interactions with the product (Karapanos, 2013). Such practices are concerned with value attributed by users during initial use, where such evaluation is assumed to represent a constant quality. Several authors have pointed out the limitations of such a flawed approach (Den Ouden *et al.*, 2006; Von Wilamowitz-Moellendorff *et al.*, 2006; Karapanos, 2013). Ever-increasing importance is being attached to longitudinal evaluation practices (Karapanos, 2013). The issue of the unstable nature of value, which Salvatierra-Garrido *et al.* (2010) referred to as dynamism, evokes a design solution that is closely linked to lean thinking, namely the flexibility of spaces. Flexibility makes an indeterminate life-period product (such as a building) evolutionarily possible, where changes can be made to adjust the right attributes for a better solution in different situations.

4.3.3 The components

Despite the argument for a systemic approach to the concept of value, it is important to observe that evidence suggests the existence of different components or aspects that define the construct of value. Note these are different from categories. In a widely cited paper, Sweeney and Soutar (2001) review and suggest a construct for value, based on a number of these components. Dimensions emerged, which were termed 'emotional value', 'social value', 'quality' or 'performance', and 'price' or 'value for money'. Sheth *et al.* (1991) propose different dimensions: social, emotional, functional, epistemic and conditional. Unlike Adam Smith, Aristotle and some lean construction authors, however, these researchers do not distinguish between different kinds of value, but rather consider a multitude of aspects to be relevant to the construction of value. Evidence suggests that some dimensions of value are more influential than others, depending on the situation (Sheth *et al.*, 1991).

4.3.4 Context and framing

Framing is known to be a key component of how people perceive and evaluate problems relating to choice (Tversky & Kahneman, 1986). Köhler (1938) devotes special attention to the importance of context in the vector settings of value, once the properties of a unit can be identified with phenomenology from the position of the object in the system. Likewise, people cannot assume that products have measurable value, except in the relation that they have with other valuable objects. Even economic goods are only valuable due to a system that embraces them.

In architecture, under the strong influence of modern historiography, Alois Riegl was the first person to identify the importance of context in evaluative issues concerning buildings, in 1903. The Viennese art historian, particularly interested in historical buildings, offers a definition of the concept of a 'historical monument'. These buildings, constructed with ordinary interpretations or purposes in previous eras, acquire monument status when they respond to new specific values in contemporary contexts (Riegl, 1903). As François Choay (1992) confirms, these values are pertinent to contemporary times, not having been observed in other historical periods.

4.3.5 The subjectivity matter

Throughout the history of philosophy, there has been considerable divergence between the objective and the subjective perspectives of value. In the modern world, the subjective notion of goods is conveyed by Hobbes, who asserts that value depends on the judgement of others. As an illustration, he explains that 'the value of a skilled military commander is high in time of war, present or imminent, but not at peacetime' (Hobbes, as cited in

Abbagnano, 1961: 989). It is, however, since Nietzsche that subjective value gains a fundamental emphasis in philosophy, when there is no value that is not a possibility, or a way of being, of man; this thesis of the interpretation of value is called 'empirical' or 'subjectivist'. Under this conception, with the rise of historicism, relativism of values is born, according to which history sets values, ideas and meanings, without them ever being an objective entity, but rather only existent in the subject–object relation. In this sense, 'there are no absolute values, and the values are just those which, under certain conditions, men recognize as such' (Abbagnano, 1961: 992).

However, there is still great confusion between the terms 'objectivity' and 'subjectivity', and elucidation of these terms is not an objective of this work. However, people should be clear about our position in defending the concept of value as issuing from the subject. Obviously, objects experience an extraneous existence to the subject, and have properties of their own – or attributes – such as the quality of 'red'. These attributes can be immutable, an invariable stimulus before the evaluation of individuals. Yet the subject does not have direct experience with these objective properties, but rather with phenomena corresponding to them, conformed in the subject itself, through complex mental processes related to perception, cognition and consciousness. Therefore, what people accept as objective observations are conscious phenomenological experiences that are very consistent both within and across subjects (Dehaene & Naccache, 2001). Subjective experiences, on the other hand, are the ones not consistent, which vary depending on personal idiosyncrasies.

Accordingly, this chapter opines that an object, whose existence is independent of the presence of the subject, only becomes an object of value when it is in relation with the subject; it becomes an object of value in our interpretations of it, and in the judgements that we make of it. Economists Menger, Jevons and Marshall agreed that individuals have personal desire scales of an ordinal nature, and that this is reflected in prices and choices (Feijó, 2010). They also understood that these units cannot be compared in different people, or at different times. Accordingly, this chapter stresses the importance of a very clear distinction between values for each of the actors in whichever value-focused relationship. For instance, a study of interest to designers and construction professionals suggests that 'architects as a group cannot predict the public's aesthetic evaluations of architecture' (Brown & Gifford, 2001: 93).

It must also be pointed out that defenders of subjectivity have often turned it into solipsism, thereby ignoring the determining influence of context and society (Vasconcellos, 2002). Because of the fragility of these concepts, some authors, such as Rooke *et al.* (2010), have even been increasingly refusing the conventional notions of objectivity and subjectivity. Accordingly, to support the dichotomy between objectivity and subjectivity is not the purpose of this chapter, as mentioned earlier, but rather to present its historical relevance, and to highlight value as a relation established between subject and object.

4.3.6 The social construction

Hubbard (1996) reveals the existence of a dichotomy between individualistic and social interpretations of environmental preferences. The author, however, defends the relevance of both interpretations in understanding the relation between the subject and space. This chapter cannot defend a value that is generated solely by the individual, and is separate from the external context, particularly from social relations. Some authors have even come to define value as a measure of a product's worth in a particular social context (Baldwin & Clark, 2000). The perception in this chapter, subordinated to meaning and value issues, also depends on conventions, on a learning related to culture. Rooke *et al.* (2010) propose the notion of intersubjectivity, rather than subjectivity, which is consistent with the properties of value evoked throughout this chapter.

The social characteristics of value are a particularly relevant object of study for economists when examining macro-economic trends and price stability. It is observed that requirements within social groups are more stable – although they do vary over time – due to a constant process of feedback and mutual anchoring. When studying value in relation to prices, Ariely *et al.* (2006) assume that anchoring, functioning as the public parameters of the economy itself, is responsible for prices and demand stability, which are observed in spite of the lack of stable personal values for ordinary products. Such collective anchoring, in turn, is triggered by historical accidents or manipulations.

In the findings by Ariely and colleagues (2006), interesting empirical evidences regarding the relationship between the concept of value and related notions begin to emerge. According to the authors, requirements and values are not so strongly related and may behave differently in similar social contexts. In the existing literature dedicated to the concept of value, from different disciplines, it is striking the focus on the relationship between value and related instrumental concepts, such as choice. As a growing body of literature suggests their dissociation, it becomes evident that a comprehensive understanding of the concept of value cannot be achieved without investigating not only the way value is constructed, but also the way value acts on the world and in society.

4.4 Practical sign of value: choices and declared preferences

Typically, values are not directly perceived. They manifest themselves through preferences and choices, thus acting on the world indirectly. Most of the importance attributed to value is based exactly on the belief that value is a determinant in behaviour, through choices, and in satisfaction and wellbeing, through fulfilment of valued requirements. As previously mentioned, however, these notions have complex behaviour and must be discussed in more depth. The tangible properties of value, as well as their forms of manifestation, must be a relevant component of understanding of the concept of value in lean construction. The next paragraphs provide an overview of a few aspects pertinent to the problem of choice and preferences in relation to value. In philosophy, Abbagnano (1961) states that value defines the preferable (choice possibility), being the object of a normative expectation. Value is the judgement criteria of choices and, although it may not always be a standard that is followed, it cannot be set aside by preferences. Value is, therefore, expected to be an important determining factor of choices and preferences.

By focusing on value, lean construction intends to meet clients' preferences, thereby generating satisfaction. This logic assumes that human beings are perfectly rational beings who always act and choose based on their best interests. This assumption, however, has been repeatedly disproved in research. What has been observed in the study of decision-making mechanisms is that individuals are not as skilled at these processes. In fact, researchers such as Iyengar and Lepper (1999) have demonstrated that brains do not deal very well with choices. Haase and Rothe-Neves (2007: 121) claim that individuals in various circumstances do not make rational decisions. They argue that judgement and decision-making processes are strongly determined by the architecture of the cognitive system: in general, it is more adaptive to make decisions based on heuristics than from a deliberate effort of rational judgement. By decisions based on heuristics is meant intuitive, rapid, evolutionarily shaped solutions.

Haase and Rothe-Neves (2007) described several heuristics with the potential to skew decision-making processes. The main ones are availability, representativeness, anchoring and emotional projection. The availability heuristic suggests that people make decisions based on the most salient information, without systematically considering other alternatives. The representativeness heuristic occurs when a person disregards the prevalence of certain conditions of the population in the judgements of their probability a posteriori. The anchoring heuristic occurs when the choice of a first reference point (to facilitate judgements under uncertainty) does not undergo the necessary adjustments, so that the anchor affects the individual's judgement. Finally, the heuristic of emotional projection is reflected in the strategy of projecting one's own present emotional state onto someone else, or onto a future state. Several other forms of bias and heuristics have been covered in the literature.

Furthermore, lean construction understands that value to the client is the variable that defines the visibility and quality of certain product attributes. What has been confirmed, however, is that the assignment of value itself is also not a direct activity, and it is subject to various determining factors with regard to perceptions. This condition has led researchers to coin the term 'perceived value'. Examples of such factors are: the speed with which a variable changes (whether it 'startles' the reasoning powers), how the information is presented and the possibilities for comparison with other options (Ariely *et al.*, 2003).

Ariely *et al.* (2003) argue that there is greater behavioural impact when people are aware of differences or changes than when they are only aware of the prevailing levels at a particular point in time. They state that 'it is more difficult to evaluate attributes separately than jointly, and that the difficulty with absolute evaluations is larger for attributes that do not have well-established standards' (Ariely *et al.*, 2003: 8). These authors also present findings from studies that show that even if one choice does not represent a preference (as when the individual does not have a preference), this choice tends to repeat itself. These studies indicate that choices do not necessarily reveal real preferences, that actions taken upon revealed preferences may not be the most appropriate actions, and that 'market institutions that maximize consumer sovereignty need to not maximize consumer welfare' (Ariely *et al.*, 2003: 45). Ariely *et al.* (2003: 45) suggest, however, that ordinal utility may be a valid representation of choices under specific circumstances: people respond coherently when differences are evident. The author concludes that 'although the distinction between "revealed" and true preferences goes against the grain of modern economic theory, it is forced on us here by experimental evidence' (Ariely *et al.*, 2003: 12).

Based on these arguments, the chapter is led to challenge the idea that responding to people's values is consistently related to increasing their welfare.

4.5 Reaching an internally consistent framework of value

Both fundamental and instrumental characteristics of value have been discussed. Although the available literature on the concept of value is more extensive than could be covered in this chapter, it is believed that the aspects of value discussed so far are sufficient to provide an initial understanding of the complex nature of value. Based on these aspects, comprehensive definitions for constructs relevant for lean construction discussions, such as that of generating value, can begin to emerge.

Based on what have been presented, value generation can be argued to consist of inducing the perceived possibility of fulfilment of clients' needs through a provided product or service; this process is mediated by a number of extrinsic factors (Teas & Agarwal, 2000) as well as human heuristics (Haase & Rothe-Neves, 2007). This definition, although suggestive only, exemplifies a concept which condensates the understanding of value argued in this chapter. The intention is not to summarize the aspects and characteristics of value being presented until this point, but only to demonstrate that the understanding of value constructed has the potential to assist in the establishment of an internally consistent framework for the study of value in lean construction. Furthermore, the proposed understanding of value raises important questions that must be addressed.

The next section initiates a discussion on the implications of the proposed way of addressing value to its application in the field of lean construction.

4.6 Implications of the study of value in lean construction

In academic research on lean construction, the concept of value does not seem to be devoid of complements to be qualified: the term used by marketing, namely 'value to the client' was adopted, with all its implications. This view of consumption is linked to several contemporary problems, as there would seem to be a patent necessity for consumption in today's society, as has been established through several instruments, such as 'planned obsolescence'.

Co-responsible for the environmental crisis on the planet, the need for consumption is just one macro example of how manipulation of value and necessity issues may have negative effects collectively. When it comes to buildings, manipulation of value and needs is even more relevant. Buildings not only include the majority of human activities, but are also responsible for the conformation of entire cities. Buildings, for instance, account for up to 40 per cent of a country's energy consumption (Pérez-Lombard *et al.*, 2008), which highlights the impact of buildings on the planet and everyday life. This special quality that the built environment acquires, and the extent of its interference at both individual and collective levels of society, leads this chapter to consider issues of responsibility when dealing with value and satisfaction in construction.

For expanding the discussion on responsibility, it is crucial to consider the result of an unlikely perfect evaluation, where the subject considers, in detail and free of heuristics, all relevant aspects regarding the object of analysis. The value that would be established from this relationship should be the value that is pursued by designers, but it is not. In both the manufacturing and construction sectors a proposal of attributes has been observed whose goal is to 'deceive' the client's perception, and, consequently, their evaluation. These attributes are often found among the so-called 'selling arguments'. From this perspective, they could be considered waste, even as objects of declared preference, since they do not bring actual benefits to the client, as an individual or as a member of society. Based on the understanding of value, certain conflicts inside the idea of defining waste based on people's values start to appear.

Menger warns that 'the belief that something has the power of wish-fulfillment may even be wrong and it still is, in fact, economic good' (Feijó, 2001: 387). In medicine, where the consequences of personal choices are taken more seriously, it has developed a wide range of bioethical debate about the so-called 'principle of decision-making autonomy' and 'informed consent'. The purpose of informed consent is to provide all necessary information for the patients, to allow them to decide according to their values (Haase & Rothe-Neves, 2007). Meanwhile, the fragility of the principle of autonomy has been questioned, for reasons that have been mentioned in this text. Given the understanding of the importance of the built environment in human life, it raises the question of whether a discussion of value in lean construction could proceed without an ethical discussion as a component.

People may argue that just as preference matters do not necessarily reflect a real preference, value – despite being a key part of client satisfaction – is not an absolute and unquestionable requirement-generating unit. On issues of environmental and historical heritage, for instance, real requirement and demand must be constantly frustrated. For example, valuation of personal interests over common interests represents an explicit issue in world conferences on sustainability, which leads to the question 'How can we work with sustainable design attributes if client demands are not always committed to sustainability?' The solution to this problem may lie in being able to create new needs in the client, as an individual or society, and thereby to create value. As Abidin and Pasquire (2005: 176) suggest, 'clients may not have the adequate knowledge to drive them into demanding sustainability'. Sustainability knowledge would help people to understand the importance of achieving sustainability in building projects, and to address issues related to the environment and social demands.

Note that the prospect of global repercussions of environmental irresponsibility, as well as their massive coverage in the media, has already made sustainability a field of potential value creation, which has been exploited by industry. But this clearly is not enough. Nor is the question always this simple. In the end, the world is caught in an old sociological debate regarding value: where does the limit lie to intervening in the evaluative behaviour of a group based on our own set of values? Even with collective and individual welfare, it can still be argued that the desirability of every action is subjective and based on personal values. If people are always evaluating the world through personal lenses, what right does a third party have to try to shape the lenses of others?

4.7 Summary

This chapter has provided a framework for the concept of value in lean construction from a systemic approach. Understanding value as a concept first, rather than as an operational definition, or as a practical tool, has raised several important questions, which have been left unanswered. The purpose is to introduce the discussion and highlight the complex implications that the study of value presents. The chapter argues that it is essential that these implications be explored and debated. This work reinforces the idea that despite close relations, lean construction and value generation alone are not sufficient to promote social and individual welfare, or responsible and sustainable development. Researchers and practitioners must understand the potential and the pitfalls of the use of value as a main indicator of waste in lean construction, so that stakeholders can harness the best outcomes of value. Despite the words of caution, the authors of this chapter opine that introduction of the concept of value through lean construction and TFV theory has been immensely important to the body of knowledge of construction and architecture. Thinking of design, production and construction processes in terms of value expands the perception of these processes, acknowledging critical aspects which have previously been overlooked. It also resonates with efforts which are currently being made in several other disciplines, which should continue to enrich in a systemic way the understanding of value.

Notes

In developing this chapter, the authors have drawn on Maia *et al.* (2011). The authors gratefully recognise the International Group for Lean Construction in this regard.

Endnotes

1   Menger was an Austrian economist who focused on the essences of concepts such as value precluding the application of mathematical methods.
2   According to LT, a task accomplished with 'maximum efficiency' or an attribute delivered with 'optimum performance', for instance, can still be considered waste if it does not compete to add value to the building.
3   The term 'complexity' used in this text refers to the concept adopted by Edgar Morin (2007). The reader should refer to this reference for an in-depth discussion of the term.
4   Menger also considers that human desires do not need to be rational, and, indeed, with the progress of civilisation, irrational desires have gradually become more important.

References

Abbagnano, N. (1961), *Dizionario di filosofia*, UTET, Torino.
Abidin, N.Z. & Pasquire, C.L. (2005), Delivering sustainability through value management: Concept and performance overview. *Engineering, Construction and Architectural Management*, Vol. 12(2), pp. 168–180.
Ariely, D., Loewenstein, G. & Prelec, D. (2003), Coherent arbitrariness: Stable demand curves without stable preferences, *The Quarterly Journal of Economics*, Vol. 118(1), pp. 73–105.
Ariely, D., Loewenstein, G. & Prelec, D. (2006), Tom Sawyer and the construction of value. *Journal of Economic Behavior & Organization*, Vol. 60(1), pp. 1–10.
Baldwin, C.Y. & Clark, K.B. (2000), *Design Rules*, MIT Press, Cambridge, MA.
Brown, G., & Gifford, R. (2001), Architects predict lay evaluations of large contemporary buildings: Whose conceptual properties? *Journal of Environmental Psychology*, Vol. 21(1), pp. 93–99.
Choay, F. (1992), *L'Allégorie du patrimoine*, Editions du Seuil, Paris.

Davis, J.B. (2013), *The Theory of the Individual in Economics: Identity and Value*, Routledge, London.

Dehaene, S. & Naccache, L. (2001), Towards a cognitive neuroscience of consciousness: Basic evidence and a workspace framework. *Cognition*, Vol. 79(1), pp. 1–37.

Den Ouden, E. (2006), Development of a design analysis model for consumer complaints. PhD thesis, Eindhoven University of Technology.

Erikshammar, J., Björnfot, A. & Gardelli, V. (2010), The ambiguity of value. In: *Proceedings of the 18th International Group of Lean Construction Conference*, Technion, Haifa, Israel, 14–16 July.

Feijó, R. (2001), *História do pensamento econômico*, Atlas, São Paulo.

Fischhoff, B. (1991), Value elicitation: Is there anything in there? *American Psychologist*, Vol. 46, pp. 835–847.

Garcia, M.J. (2003), Em busca do conceito de valor, *Cadernos de Semiótica Aplicada*, Vol. 1(2), pp. 59–108.

Haase, V.G., Pinheiro Chagas, P. & Rothe-Neves, R. (2007), Neuropsicologia e autonomia decisória: Implicações para o consentimento informado. *Revista Bioética*, Vol. 15, pp. 117–132.

Hubbard, P. (1996), Conflicting interpretations of architecture: An empirical investigation. *Journal of Environmental Psychology*, Vol. 16, pp. 75–92.

Iyengar, S.S. & Lepper, M.R. (1999), Rethinking the value of choice: A cultural perspective on intrinsic motivation. *Journal of Personality and Social Psychology*, Vol. 76(3), pp. 349–366.

Karapanos, E. (2013), User experience over time. In: *Modeling Users' Experiences with Interactive Systems*, Springer, Berlin, Heidelberg, pp. 57–83.

Köhler, W. (1938), *The Place of Value in a World of Facts*, Liveright, New York.

Koskela, L. (2000), *An Exploration Towards a Production Theory and its Application to Construction*, VTT Building Technology, Espoo.

Lima, M., Maia, S. & Barros Neto, J.P. (2011), Epistemological basis of design process: Positivism versus complexity. In: *Proceedings of the 19th Annual Conference on Lean Construction*, IGLC, Lima, Peru, 13–15 July.

Maia, S., Lima, M. & Barros Neto, J. (2011), TA systemic approach to the concept of value and its effects on lean construction. In: *Proceedings of the 19th Annual Conference on Lean Construction*, IGLC, Lima, Peru, 13–15 July, pp. 22–31.

Menger, C. (2007), *Principles of Economics*, Mises Institute, Auburn, AL.

Morin, E. (2007), Restricted complexity, general complexity. In: C. Gershenson, D. Aerts & B. Edmonds (eds), *Worldviews, Science and Us: Philosophy and Complexity*, World Scientific, Singapore, pp. 5–29.

Pérez-Lombard, L., Ortiz, J. & Pout, C. (2008), A review on buildings energy consumption information. *Energy and Buildings*, Vol. 40(3), pp. 394–398.

Riegl, A. (1903), *Der moderne Denkmalkultus, sein Wesen, seine Entstehung*, Kessinger Publishing, Vienna.

Rittel, H. (1966), Überlegungen zur wissenschaftlichen und politischen Bedeutung der Entscheidungstheorien. In: H. Krauch, W. Kunz & H. Rittel (eds), *Forschungsplanung*, Oldenbourg Verlag, Munich, pp. 110–129.

Rooke, J., Sapountzis, S., Koskela, L., Codinhoto, R. & Kagioglou, M. (2010), Lean knowledge management: The problem of value. In: *Proceedings of the 18th International Group of Lean Construction Conference*, Technion, Haifa, Israel, 14–16 July.

Salvatierra-Garrido, J., Pasquire, C. & Thorpe, T. (2010), Critical review of the concept of value in lean construction theory. In: *Proceedings of the 18th International Group of Lean Construction Conference*, Technion, Haifa, Israel, 14–6 July.

Sheth, J.N., Bruce, I.N. & Barbara, L.G. (1991), Why we buy what we buy: A theory of consumption values. *Journal of Business Research*, Vol. 22, pp. 159–170.

Sweeney, J.C. & Soutar, G.N. (2001), Consumer perceived value: The development of a multiple item scale. *Journal of Retailing*, Vol. 77(2), pp. 203–220.

Teas, R.K. & Agarwal, S. (2000), The effects of extrinsic product cues on consumers' perceptions of quality, sacrifice, and value. *Journal of the Academy of Marketing Science*, Vol. 28(2), pp. 278–290.

Thyssen, M.H., Emmitt, S., Bonke, S. & Kirk-Christoffersen, A. (2010), Facilitating client value creation in the conceptual design phase of construction projects: a workshop approach. *Architectural Engineering and Design Management*, Vol. 6, pp. 18–30.

Tversky, A. & Kahneman, D. (1986), Rational choice and the framing of decisions. *The Journal of Business*, Vol. 59(4), pp. 251–278.

Vasconcellos, M.J.E. (2002), *Pensamento sistêmico: O novo paradigma da ciência*, Campinas: Papirus.

Von Wilamowitz-Moellendorff, M., Hassenzahl, M. & Platz, A. (2006), Dynamics of user experience: How the perceived quality of mobile phones changes over time. In: *Proceedings of the 4th Nordic Conference on Human–Computer Interaction*, pp. 74–78.

Woodruff, R.B. (1997), Customer value: The next source for competitive advantage. *Journal of the Academy of Marketing Science*, Vol. 25(2), pp. 139–153.

5   Value is relative

How decision-making theories affect lean construction

Bolivar A. Senior

This chapter presents a critical review of decision-making theory aspects of relevance to lean construction. It identifies natural human tendencies concerning decision making that can distort rational outcomes and the lean construction features that could be impacted by these biases. Using vocabulary from decision-making theory, lean construction can be described as an enriched option relative to traditional management approaches. Enriched options lead to paradoxical, simultaneous stronger reactions for their adoption and for their rejection, depending on the framing used for their discussion. Lean construction techniques such as phasing scheduling in the Last Planner System™ can be impacted by subtle factors in the manner in which they are introduced to a new audience. Moreover, technical considerations can be influenced by psychological factors. Examples include the possibility of delaying the last responsible moment of a decision if the decision must be made from many similar choices and the altogether rejection of lean construction if it is perceived as entailing too many features that do not add value to a company's current needs. Such behavioural examples point to the important role of psychology in the creation and management of value for lean construction.

5.1 Background

An implicit assumption for the introduction of lean construction (LC) to construction companies using a traditional management approach is that if an LC method such as the Last Planner System (LPS) is a superior option compared to the methods currently used by a company, and provided that these advantages are clearly explained to the company, the company would choose LC over its current management approach. But, as discussed here, the selection process is significantly complicated by natural human attitudes towards competing alternatives and change. LC is a relatively new management approach, and as such it is particularly affected by its role as a challenger to established views and procedures.

All humans, individually and in groups, constantly need to make decisions. As Tannenbaum (1964) points out, the need for making decisions arises 'out of the fact that knowledge of relevant existing facts is inadequate and that the future is uncertain'. Every decision relies to some extent on assumptions that lead to selecting one choice over its alternatives. These assumptions fill in the inevitable holes arising from imperfect information and outcome uncertainty. Lean thinking (Womack, 1996) provides comparatively 'more autonomy in production decisions and enriched jobs as a consequence of the lean principles regarding distributed decision-making, multi-skilling, and pursuit of perfection' (Howell & Ballard, 1999). LC has adapted lean principles to the construction industry (Koskela, 1992). It considers that many project planning and execution decisions should be made by field managers, since these individuals are in the best position to understand the decision at hand (Howell & Ballard, 1998). A construction project has many possible alternatives for action at any given point. LC provides guidelines for these required decisions, but in the final analysis, each one is the outcome of human thinking only informed by these recommendations. The central role of decision making in LC management makes imperative the understanding of the decision-making mechanisms underlying the actions required to complete a construction project.

This chapter begins with a broad-level discussion of historical and current decision-making paradigms. Each paradigm frames a set of theories attempting to explain the motivation and tendencies in the process of making decisions. More specific aspects of decision-making are addressed next, with special attention to the issues that can affect the introduction and implementation of LC. Lastly, this chapter deliberates on the implications of its findings to the concept of value and the implementation of LC.

5.2 Decision-making paradigms

Several theories, models and paradigms have been put forward to explain human behaviour when choosing among alternatives. This study groups these explanations into two distinct frames. The value maximisation paradigm begins with the assumption that decisions are based on the human desire to maximise the value offered by the chosen alternative. Theories in this category assume that human beings act rationally, and offer a quasi-mathematical explanation and optimisation of the decision-making process. The intuitive reasoning paradigm groups theories based on evidence strongly suggesting that humans are influenced by factors more complex than the rational behaviour assumed by the value maximisation paradigm. These factors can be seemingly irrelevant to the decision at hand and lead to irrational choices that are nevertheless consistent and predictable. The central assumption of the intuitive reasoning paradigm is, paraphrasing Ariely (2008), that human behaviour in decision making can be predictably irrational.

5.2.1 Value maximisation paradigm

The value maximisation paradigm is based on the concept of expected utility. It proposes that a decision maker will choose the option that maximises the weighted sums obtained by adding the utility values of outcomes multiplied by their respective probabilities (Levy, 1992). They have been the basis for many practical applications such as advertising strategies and financial forecasting.

Von Neumann–Morgenstern theory. The most mathematically oriented of theories under the value maximisation paradigm is the Von Neumann–Morgenstern Theory (VNMT). It was introduced in 1944 as a mathematical theorem examining utility value behaviour under risk, i.e. under known factors subject to probability in their outcomes (Von Neumann & Morgenstern, 1944). In the VNMT, a person (or 'agent') is rational if and only if his or her behaviour maximises the expected value of the set of possible outcomes. To achieve this, an agent needs to define measures of risk and value which in practice are extremely difficult to quantify (Dyer & Jia, 1998). VNMT is exemplary of a Descartian view of the world, in which human beings are capable of totally rational decisions such as planning their future (Senior, 2007). VNMT brought decision making to the forefront of the field of economics, and is relevant as a reference point for the evolution of decision-making theories.

Prospect theory. Decision-making theory was significantly advanced by prospect theory. It follows three principles summarised by Kahneman and Tversky (1979) as follows. (1) Expectation: the overall utility of a prospect is the expected utility of its outcomes. (2) Asset integration: a prospect is acceptable if the utility resulting from integrating the prospect with one's assets exceeds the utility of those assets alone. (3) Risk aversion: people are generally risk averse. This means that most people will prefer an alternative with expected value *x* over any riskier alternative with equal expected value *x*. The curve in Figure 5.1 shows a value function plotting the value assigned by an average person to the various outcomes resulting from a given choice.

The figure shows that perceived value does not increase with gain as quickly as it decreases with loss. While perceived value tends to flatten after a certain gain is reached, the value of the outcome keeps decreasing as the loss increases. An extra $1,000 are more valuable to an average-income person than to a millionaire. And a millionaire will lament the loss of $1,000 dollars more than he or she will value the gaining of the same amount.

5.2.2 Intuitive reasoning paradigm

The imperfect information and uncertainty in outcome of all alternatives leading to a decision make inevitable some subjective reasoning in the process of arriving at the decision. As soon as subjective reasoning is involved in the decision, purely rational decision making is difficult to apply

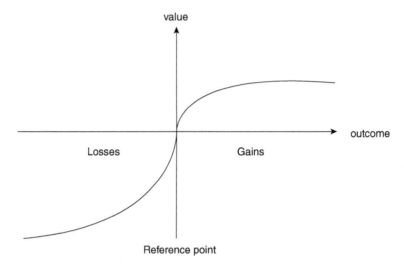

Figure 5.1 Prospect theory value function

(Time for Change, 2012). In fact, many studies 'leave no doubt about the failure of rational choice as a descriptive model of human behaviour' (Jones, 1999). The intuitive reasoning paradigm frames substantially more complex processes in which intuition plays a central role. Intuition has been defined as 'a non-sequential information-processing mode, which comprises both cognitive and affective elements and results in direct knowing without any use of conscious reasoning' (Sinclair, 2005). The intuitive reasoning paradigm requires an understanding of human behaviour at levels that are difficult or impossible to rationalise.

Bounded rationality. A well-known attempt to develop a theory beyond the value maximisation paradigm is the bounded rationality theory of Herbert Simon (Simon, 1991). Bounded rationality considers that people act rationally under the limitations of both their knowledge and their cognitive capacity. This theory introduced some concepts increasingly accepted in decision-making theory and in economics, such as that people tend to use heuristics (educated rules of thumb) to reach decisions, and that in many instances the objective of decisions is the satisficing of needs (this word combining 'satisfying' and 'sufficing') instead of the absolute optimum assumed as the objective of theories in the value maximisation paradigm. Notable contributions of bounded rationality theory include recognising the costs of gathering and processing information. These costs can have a significant effect on the value assigned to an alternative. Bounded rationality and other closely related derived theories have had an immense influence in current decision-making theory. However, some researchers find their approach insufficient to explain the decision-making process. For example, Etzioni (2011) complains that bounded rationality does not explain satisfactorily the irrationality of many decision-making situations, and that instead it lowers the bar 'by defining down that which is entailed by being rational'.

Behavioural economics. A loose continuum of explanations for decision making in which psychological factors play a central role has been developed under the umbrella of behavioural economics. Etzioni (2011) provides a view of the underpinnings of this theory:

> [Research studies] show beyond reasonable doubt that: (a) Indeed, much choice behaviour is not based on deliberations of any kind; (b) when reasoning does occur, it is often subject to the cognitive biases B.E. [behavioural economics] systematically observed and reported; and (c) both 'intuitive' … choices and those subject to deliberations are deeply affected by emotions and norms, and these in turn by social and cultural factors.

A cognitive bias is any systematic deviation from a normative criterion that affects thinking, often leading to errors in judgement (Han & Lerner, 2009). It can be both unintentional and unconscious (Hamilton Krieger, 1995). Behavioural economics includes well-known economists, such as Nobel Prize winner Kahneman, popular authors (e.g. Ariely, 2008; Gladwell, 2000) and researchers (e.g. Etzioni, 2011; Tversky, 1974). Many recent studies have concentrated on the intensely psychological emphasis of this theory. The relevant aspects of decision making discussed in the next section come from behavioural economics experiments.

5.3 Relevant aspects of decision making

5.3.1 Dominance and conflict

A condition of dominance arises when an alternative is perceived as superior to another in all significant features. In contrast, a conflict condition arises when one alternative may be superior to another in only some dimensions (Shafir *et al.*, 1993). LC, for example, would be dominant compared to traditional management techniques if the former is perceived to be superior to the latter in all significant aspects.

Conflict conditions complicate decision making. Experiments have shown that opting to search for additional alternatives does not depend only on the value of the best alternative, but the level of conflict among the alternatives already considered. When options are in conflict, more alternatives may be sought, the decision to choose one may be postponed or the existing options may be subject to further scrutiny with the possibility of cognitive biases such as the ones described in the following subsections.

5.3.2 Enriched and impoverished options

If one decision option has both more positive and more negative features than another one, then the former is an enriched option compared to the latter. The latter would be an impoverished option relative to the former.

Individuals and groups choosing between two courses of action tend to select the one with most prominent positive features and reject the one that has the most prominent negative features (Slovic, 1975; Shafir, 1993). An enriched option is, paradoxically, more likely to be selected and also more likely to be rejected than an impoverished option, depending on how the decision is framed.

Suppose that a construction company needs to choose between two productivity improvement techniques, A and B.

- A significantly improves the construction production rate and eliminates the need for the majority of change orders. It requires major changes to the company's management practices and continuous commitment by all employees.
- B provides some improvement to production rate and eliminates some change orders. It requires small changes to the current management practices and requires employees to attend a single training seminar.

If the situation is framed as 'Which of the two choices should the company keep?', then A is more likely to be kept because of its advantages over B. But, if the question is phrased as 'Which one should the company discard?', then experimental findings indicate that A is likely to be discarded because of its disadvantages compared to B. The phrasing of whether to keep or discard the enriched choice leads to contradictory results. The above scenario is plausible if LC is weighted against management procedures already in place in a construction company. A successful lean implementation requires company commitment and challenges to traditional management structures that could be viewed as negative factors by some individuals. The rewards of a successful lean implementation are many. LC is an enriched option in this scenario, and a careful structuring of the wording used throughout its implementation (the framing of its implementation) can lead to success or failure.

5.3.3 Independence of irrelevant alternatives

The principle of independence of irrelevant alternatives, also called the regularity condition of value maximization (Shafir *et al.*, 1993), states that if Option 1 is preferred over Option 2, then the addition of an Option 3 of less value than Option 2 should not make a decision maker switch preferences.

This principle frequently does not hold true when applied to practical situations. The author replicated an experiment to this effect described by

Ariely (2008). A magazine advertisement similar to the one for Scenario A in Figure 5.2 was shown to a group of students. The internet-only subscription is listed at $40, the print-only subscription at $85, and the internet and print combined subscription at $85. Students had to choose one option, i.e. there was no 'None of the above' option. No student chose the print-only alternative, 57 per cent (20 of 35) chose the internet-plus-print alternative, and the remaining 43 per cent selected the internet-only option. A second group was shown Scenario B, where the print-only version was removed. This group preferred the internet-only option by 87 per cent (27 of 31) compared to 13 per cent of the internet-plus-print option. The removal of a seemingly irrelevant third option led to a dramatic change in preferences.

5.3.4 Differing the time required for a decision

The time required for reaching a decision is affected by the number of available choices and the level of similarity among them (Shafir *et al.*, 1993). As an example, assume that subcontractors A and B bid for a job and that subcontractor A is deemed to be the best choice. The indecision introduced by the same scenario if subcontractors C and D also bid (i.e. A, B, C and D bid for the job) would lead to a disproportionally longer time to select a winner, even if C and D are clearly inferior to A. This tendency to defer choice is more pronounced when A and B are of similar perceived value (e.g. Huber *et al.*, 1982). Most individuals are averse to analysing the trade-offs required for choosing between similar options, especially when both are valuable. The addition of options, in fact, makes more appealing the choice of doing nothing if this is a possible outcome of the decision-making process.

Lean management in general defines the last responsible moment as 'the instant in which the cost of the delay of a decision surpasses the benefit of delay' (Lean Tools, 2012). Lean construction uses this principle to recommend, among other applications, that 'design decisions will be deferred until the last responsible moment if doing so offers an opportunity

Magazine Subscription Offer	

Scenario A

Internet-only annual subscription	$40
Print-only annual subscription	$85
Internet plus print annual subscription	$85

Scenario B

Internet-only annual subscription	$40
Internet plus print annual subscription	$85

Figure 5.2 Example of paradox of independence of irrelevant alternatives

to increase customer value' (Ballard, 2000b). However, this principle could have unexpected consequences, since a decision may be deferred as a consequence of the natural reaction to conflict in the presence of many similar options. There is a theoretical possibility of this tendency towards deferment resulting in postponing action for too long. This possibility does not appear to have been addressed by lean construction researchers.

5.3.5 Non-valued features

One common device intended for encouraging the choice of an alternative is to expand the offer with extra features or items. Those features may be irrelevant to the choice or even not wanted. These additional features have the purpose of increasing the attractiveness of the main offer, and indeed they may be important for some individuals. However, several studies have shown that non-valued features do not act as incentives. On the contrary, individuals are reluctant to choose alternatives loaded with (subjectively) not valued features. Simonson *et al.* (1994) found that the tendency not to choose alternatives with unwanted features holds true even when the feature is offered for free.

Should lean construction techniques such as the LPS (Ballard, 2000a) be limited, at least during their implementation stages, to their bare minimum? The question is appropriate, not only for the logistics of the initial implementation, but also for the possibility of including features that may be initially undesired.

5.4 Reflection on the concept of value

Ballard and Howell (2004) summarise the three key goals of lean construction as 'delivering the product while maximising value and minimizing waste'. For lean construction, 'the value concept focuses on matching all customer requirements in the best way possible (design and production), therefore creating value from the point of view of the customer' (Henrich *et al.*, 2007). Value is thus generally recognised as a subjective property aligned with the mental accounting of decision-making theory.

Psychological aspects of value are not fully encompassed by the definitions of value found in lean construction literature. Decision-making research has shown that a customer's perceived value of an option is more than subjective: it has unconscious and malleable dimensions which can lead to irrational decisions from a value maximisation viewpoint.

Value can be fabricated. Advertising, for example, is about creating value for a customer that did not know that he 'needed' an advertised product. It can be destroyed, as in the case of negative political campaign advertising. Value hardly has the solid (if subjective) nature that lean thinking in general seems to attribute to it.

The role of psychology in the creation, management and even manipulation of value in an LC context needs to be researched. A deeper understanding of its meaning would lead to a better definition of the role that lean construction plays in the management of customer value.

5.5 Summary

There are human tendencies that distort the outcomes of human decision making in ways that could not be considered as rational from a strict value maximisation viewpoint. The potential consequences of these biases have been discussed in this chapter, based on extrapolation of experimental results in disciplines other than construction management. At the present moment, the identified consequences for lean construction are speculative, since there is a significant lack of research on the decision making in LC necessary to validate these potential consequences.

Decision making is more complicated than a relatively simple quest for maximisation of value. Individual decisions are subject to many extraneous considerations, mostly related to the decision maker's psyche. The consideration of reasons for arriving at a decision is complicated by the fact that this process includes subjective factors such as the human tendency to avoid decisions under uncertainty, to prefer options with salient features or to reject options with features of no value to the decision maker, even if they are free. These factors are frequently hidden from a person's awareness. A decision that may seem perfectly rational to the person taking it may be inexplicable or even irrational if psychological factors are not considered. Decision-making theory frames these explanations, and therefore contributes to a better understanding of lean construction's opportunities and challenges.

Significant implications of this chapter include the following:

- LC can be described as an enriched option, with more salient features relative to traditional management approaches. Enriched options lead to stronger reactions of adoption and rejection depending on the framing used for discussing their merits.
- The addition or suppression of choice alternatives affects outcomes independently of the apparent relevance of the added or suppressed alternative. LC techniques such as the LPS may be affected by this phenomenon.
- It is not clear to what extent the important LC recommendation of deferring decisions until their last responsible moment may be affected by the presence of many similar choices. Multiple similar choices have been shown as leading to excessive deferment and excessively opting for the do-nothing option.
- Non-valued features can have negative effects on the perception of overall value towards a given option. LC techniques may be affected by

this aspect of decision making if too many features are included in its techniques.

Areas for further research on the topics addressed here have been addressed throughout this chapter, and closely follow the aspects summarised above. An additional area for further research was also mentioned in the brief discussion about the concept of value. This concept, central to lean construction, should be revisited to incorporate the psychological aspects found by studies in decision making.

Note

In developing this chapter, the author has drawn on Senior (2012). The author gratefully recognises the International Group for Lean Construction in this regard.

References

Ariely, D. (2008). *Predictably Irrational: The Hidden Forces that Shape our Decisions*, HarperCollins Publishers, New York.
Ballard, G. & Howell, G. (2004). Competing construction management paradigms. *Lean Construction Journal*, Vol. 1(1), pp. 38–45.
Ballard, G. (2000a). The last planner system of production control. Doctoral thesis, University of Birmingham.
Ballard, G. (2000b). *Lean Construction Institute: Research Agenda*. Retrieved from www.leanconstruction.org/lpds.htm.
Dyer, J.S. & Jia, J. (1998). Preference conditions for utility models: A risk–value perspective. *Annals of Operations Research*, Vol. 80, pp. 167–182.
Etzioni, A. (2011). Behavioral economics: Toward a new paradigm. *American Behavioral Scientist*, Vol. 55(8), pp. 1099–1119.
Gladwell, M. (2000). *The Tipping Point: How Little Things can Make a Big Difference*, Little Brown, New York.
Hamilton Krieger, L. (1995). The content of our categories: A cognitive bias approach to discrimination and equal employment opportunity. *Stanford Law Review*, Vol. 47(6), pp. 1161–1248.
Han, S. & Lerner, J.S. (2009). Decision making. In: D. Sander & K. Scherer (eds), *The Oxford Companion to the Affective Sciences*, Oxford University Press, New York, pp. 111–113.
Henrich, G., Bertelsen, S., Koskela, L., Kraemer, K. Rooke, J. & Owen, R. (2007). Construction physics: Understanding the flows in a construction process. In: *Proceedings of the 15th International Conference on Lean Construction*, Lansing MI, 18–20 July.
Howell, G. & Ballard, G. (1998). Implementing lean construction: Understanding and action. In: *Proceedings of the 6th International Conference on Lean Construction*, Guaruja, Brazil, 13–15 August.

Howell, G. & Ballard, G. (1999). Bringing light to the dark side of lean construction: A response to Stuart Green. In: *Proceedings of the 7th International Conference on Lean Construction*, Berkeley, CA, 26–28 July.

Huber, J., Payne, J. & Puto, C. (1982). Adding asymmetrically dominated alternatives: Violations of regularity and the similarity hypothesis. *Journal of Consumer Research*, Vol. 9, pp. 90–98.

Jones, B. (1999). Bounded rationality. *Annual Review of Political Science*, Vol. 2, pp. 297–321.

Kahneman, D. & Tversky, A. (1979). Prospect theory: An analysis of decision under risk. *Econometrica*, Vol. 47(2), pp. 263–292.

Koskela, L. (1992). Application of the new production philosophy to construction. Technical Report 72, CIFE, Stanford University.

Lean Tools (2012). The last responsible moment. Retrieved from www.dzone.com/links/lean_tools_the_last_responsible_moment.html.

Levy, J. (1992). An introduction to prospect theory. *Political Psychology*, Vol. 13(2), pp. 171–186.

Senior, B.A. (2007). Implications of action theories to lean construction applications. In: *Proceedings of the 15th International Conference on Lean Construction*, East Lansing, MI, 18–20 July.

Senior, B.A. (2012) An analysis of decision-making theories applied to lean construction. In: *Proceedings of the 20th Conference of the International Group for Lean Construction (IGLC)*, San Diego, CA, 18–20 July, pp. 31–40.

Shafir, E. (1993). Choosing versus rejecting: Why some options are both better and worse than others. *Memory and Cognition*, Vol. 21(4), pp. 546–556.

Shafir, E., Simonson, I. & Tversky, A. (1993). Reason-based choice. *Cognition*, Vol. 49, pp. 11–36.

Simon, H. (1991). Bounded rationality and organizational learning. *Organization Science*, Vol. 2(1), pp. 125–134.

Simonson, I., Carmon, Z. & O'Curry, S. (1994). Experimental evidence on the negative effect of product features and sales promotion on brand choice. *Marketing Science*, Vol. 13(1), pp. 23–40.

Sinclair, M. (2005). Intuition: myth or a decision-making tool? *Management Learning*, Vol. 36(3), pp. 353–370.

Slovic, P. (1975). Choice between equally valued alternatives. *Journal of Experimental Psychology: Human Perception and Performance*, Vol. 1, pp. 280–287.

Tannenbaum, R. (1964). Managerial decision making. In D.E. Porter & P.B. Applewhite (eds), *Studies in Organizational Behavior and Management*, International Textbooks, Scranton, PA.

Time for Change (2012). Definition of intuition: What is intuitive decision making? Retrieved from http://timeforchange.org.

Tversky, A. & Kahneman, D. (1974). Judgment under uncertainty: Heuristics and biases. *Science*, Vol. 185, pp. 1124–1131.

Von Neumann, J. & Morgenstern, O. (1944). *Theory of Games and Economic Behavior*, Princeton University Press, Princeton, NJ.

Womack, J. & Jones, D. (1996). *Lean Thinking: Banish Waste and Create Wealth in Your Corporation*, Simon & Schuster, New York.

6 Lean and sustainability in construction

Creating value

Fidelis A. Emuze, Alex Opoku and John J. Smallwood

Improvement measures often set economic gains as a priority. This trend is oblivious to the synergy that exists when reducing waste to create economic gains through lean, while at the same time realising sustainability. To debate this trend, this chapter presents the relationship between lean and sustainability in relation to value. The framework in the chapter shows that value could serve as the locus of this relationship, since a review of lean and sustainability in construction indicates enhanced overall reduction of waste, and a value paradigm that is acceptable to firms and society. For instance, health and safety (H&S) as a catalyst for creating value in terms of lean and sustainability provides a platform for continual improvement at the project level and the corporate level of a construction firm. A new value lens suggests new needs to meet, new services to offer, and new ways to configure the value chain.

6.1 Background

The nature of construction is set to change in response to the intricacies of the so-called 'green economy', which is a driver of sustainable development (Ganda & Ngwakwe, 2014). Whether it is the approach to technology and/ or to new business models, there is the assumption that things will need to be done differently in order to adapt to current trends (De Valence & Runeson, 2011). When this assumption is related to construction, it suggests that change is mandatory, and that the need to make sense of construction improvement through lean and other similar management philosophies cannot be overlooked (Green, 2011). The triad of cost, quality and time, which has served as the 'iron triangle' of performance, and the anchor for value, will need to accommodate current economic, environmental and social needs.

While there are distinct collections of work related to H&S, as well as lean and sustainability in construction, there is a paucity of research that has addressed the relationship between these strategic initiatives, particularly in relation to value. It is important to explore this relationship, so as not to miss the synergies available through improved simultaneous implementation, which ensures that important trade-offs that may arise through mismatches between the initiatives are addressed. The construction management literature refers to lean and sustainability as compatible initiatives, which have a common focus on waste reduction (Ma, 2011; Mollenkopf *et al.*, 2010). When lean enables value-added activities to flow in the production environment, reduced environmental impact manifests due to optimal logistics (Wu, 2003), and, in the process, value is created.

The concept of value has been widely cited within the lean construction research community. However, the literature shows that the concept appears to be broadly defined at the project level, with limited perspectives on the business/corporate level of an enterprise. To attempt to bridge this gap, this chapter shows that value could serve as the locus of an integration of lean and sustainability in construction. As a guide for the conceptualisation of value, the chapter endeavours to emphasise that *the dimension of value should promote lean and sustainability in construction.*

6.2 Dimensions of user value

'Value' is a polysemous word, with meanings ranging from concepts related to economic return to concepts related to moral standards (Kamakura & Novak, 1992). Value has been defined through the respective lenses of sociology, psychology, marketing and industrial design. Please see Park and Han (2013: 275) for the different definitions of value in these respective disciplines. These definitions suggest that people hold more than one value, and that these values have different levels of relevance in terms of the motivation of each person, either at the personal or the corporate level. Besides the lack of consensus with regard to the meaning of the word 'value', there appear to be four main approaches to the definition of 'value'. The approaches include (Graeber, 2001):

- the notion of values as conceptions of what is ultimately good in human life;
- value as a person's willingness to pay the price of a good in terms of a cash return for certain product benefits;
- value as meaning and meaningful difference; and
- value as action.

Among these approaches is value as a person's willingness to pay the price of a good in terms of a cash return for certain product benefits, as found within the economic and business sense of capitalism. This approach will be the focus of this chapter, since it refers to the evaluation of some object, product or service by some subject or user. This focus offers a relevant basis for a discussion of value in construction, due to its user–product/service interface, and, more specifically, clients are prepared to pay for value-adding activities (Forbes & Ahmed, 2011). In this approach, the emphasis is on exchange and money, which is seen as a fundamental index of value (Carruthers & Babb, 1996). The industrial design exposition of user value by Boztepe (2007) appears to be particularly relevant for this chapter. Boztepe (2007) asserts that user value can be understood as an exchange and use, as a sign, and as an experience. The exchange approach indicates that value arises from price and the desire of a client for the product, and, as such, value is objectively determined by price. The unit of analysis of value in this approach is the exchange situation, and the product is what the user in the form of money gives up. The implication for design and construction is to ensure the quality of finished products – for instance a building, whose qualities should be functional and visible.

The lean philosophy takes cognisance of this approach when it advocates elimination of waste and improvement of value in the flow of activities in the construction process. The idea here is to ensure that the client is completely satisfied with the product (Womack & Jones, 1994). Green (2011) presents three models of lean: waste elimination, partnering, and structuring of the context. The first model recognised that waste elimination is critical for value creation when focusing on construction operations. The aim of the model is to ensure an uninterrupted flow of scheduled activities. The model assumed that cost savings made at the project level will aggregate to the corporate level, and that all parties should benefit equally from improved performance. In other words, value should accrue to project stakeholders at various levels. Many lean research findings support the gains of this model (Forbes & Ahmed, 2011). The experience approach to value is said to derive from the interaction between the user and a product within a specific sociocultural setting (Table 6.1). Value is therefore objectively and subjectively determined in the setting. The unit of analysis is always the point of experience, which refers to the product as the element

Table 6.1 A summary of the definitional approaches to user value

Perspective	The exchange approach	The experience approach
Value arises from...	... the price of and the desire for a product.	... interaction between user and product within a particular sociocultural setting.
Value is...	... objectively determined in terms of price.	... both objectively and subjectively determined.
The unit of analysis is...	... an exchange situation.	... any point of experience with the product.
The product is...	... a sacrifice made by the user, measured in terms of money.	... an element that enables experience.
The implications for design and construction are...	... the need to make the product qualities visible and functional.	... the need to understand the make-up of an experience.

Source: adapted from Boztepe (2007: 58).

that brings about the experience. The implications for design and construction can be found in the need to understand the make-up of such an experience.

Without doubt, subjective inferences are made in the experience approach to value and sustainability. The traditional model of economic, environmental and social sustainability emerged from Brundtland's (WCED 1987) report. Sustainability in the built environment has thus sought to address these three major aspects through both objective and subjective lenses. Sustainable development that meets the needs of the present without compromising the needs of the future is at the centre of industrial and academic discourse, as a result of the impact of human activities on the environment (Kibert, 2009). From a technical point of view, the experiences of sustainability, through either 'green' buildings or 'green' construction, are driven by a range of issues, such as management, specifically overall management policy, the commissioning of site management, and procedural issues:

- energy use: operational energy and carbon dioxide (CO_2) issues;
- health and well-being: indoor and external issues affecting health and well-being;
- pollution: air and water pollution issues;
- transport: transport-related CO_2 and location-related factors;
- land use: greenfield and brownfield sites;
- ecology: the ecological value of conservation and enhancement of the site;
- materials: the environmental implications of building materials, including life-cycle impacts; and
- water: consumption and water efficiency (Warnock, 2007; Shen *et al.*, 2008; Matar *et al.*, 2008).

The issues influence the determination of what is value when building sustainably. Here, lean construction principles and sustainability ideas could come together to determine superior user value in the construction context. The unit of analysis is the project, which is to be delivered to the satisfaction of the client.

6.3 Dimensions of shared value

The economic aspect of value is relevant to shared value. The literature has shown that macroeconomic trends are driving firms to create sustainable business models built on the traditional model of sustainability. Shared value encourages firms to create value in a way that goes beyond short-term economic gains, by considering a broad set of factors that determine the long-term success of the business. The objective of the way of thinking about shared value is to optimise value for the firm, and the larger society in which the firm operates (Porter & Kramer, 2006, 2011). It entails creating economic value simultaneously with social value, by addressing needs and challenges. It goes beyond corporate social responsibility, and includes sustainability in the quest for economic success. Reconceiving of the interface between society and corporate performance creates shared value (Porter & Kramer, 2011).

Table 6.2 provides a summary of the concept of shared value. It is important to note that the concept of shared value began with the realisation of the fact that organisations may have been creating value narrowly, by optimising short-term financial performance, and neglecting broader influences that determine their long-term success. According to Porter and Kramer (2011), the concept of shared value acknowledges that social needs, as opposed to conventional economic needs, define markets. They note that the concept recognises that social weaknesses often create internal costs for firms, in the form of wasted energy and costly accidents. As shown in Table 6.2, shared value can evolve when firms, such as construction firms, reconceive their products and services within the context of their location in the market, or sector. When firms also engage in exercises that redefine their perception of productivity in the value chain, shared value should result. As an illustration, when contractors follow a transformation–flow–value (TFV) perspective in their processes, production waste can be eliminated, and productivity can be enhanced (Koskela *et al.*, 2013).

Another important aspect in shared value creation pertains to the development of local clusters that have an impact on the business of the firm. The concepts of lean supply chain management could address this aspect, but it must go beyond the immediate project stakeholders, so that collaboration is entered into with institutions in the locality in which the firm operates (for detailed descriptions of shared value creation, see Porter and Kramer (2011)). The exposition by Maltz and Schein (2013), which refers to the work of Porter and Kramer (2011), among others, shows that the shared value perspective complements the concepts of sustainability, as it adopts the long-term view of value creation articulated by the 1987 Brundtland report. The perspective contends that, among other things:

Table 6.2 A summary of the concept of shared value

Concept	Creation	Practice
The concept of shared value focuses on the connection between social and economic progress. The concept has the potential to promote substantial business growth.	Shared value can be created by: • reforming product and market ideas; • redefining productivity in the value chain; and • enabling local cluster development.	Every firm needs to examine decisions and opportunities through the lens of shared value. Shared value will lead to new approaches that stimulate increased innovation and growth for firms and the society.

Source: adapted from Porter & Kramer (2011: 67).

- shared value explicitly considers value creation as more than merely the value that accrues to a firm;
- the competitiveness of a firm, and the health of the communities surrounding the firm, are closely intertwined; and
- shared value is not concerned with the redistribution of existing value, but rather is concerned with finding ways to leverage the connections between social and economic progress, so as to create additional value, which is then shared among multiple stakeholders.

Once created, shared value has to be continually optimised by the firm. Maltz and Schein (2013) observe that shared value can be optimised by cultivating the supply chain, and through collaborative, research and development capabilities. The framework proposed by Maltz and Schein (2013) shows that a relative emphasis on social value by a firm in a consistent manner will work together to leverage capabilities and their cultivation, so as to produce shared value. To start with, in construction, value propositions should be established at the corporate level, where shared value should be promoted. When the descriptions in Porter and Kramer (2011) are juxtaposed with concepts of lean and sustainability in the built environment, the need for leaders and managers in the industry to develop new skills and knowledge becomes clear. The new know-how should assist stakeholders in construction firms to grasp and/or explore the emerging approach to developing sustainable business practices, namely shared value.

6.4 Conceptual implications for construction management

The above brief review of existing management literature related to the dimensions of value shows implications for the practice of construction management, particularly in terms of a locus, or loci, for shifting the paradigm. While recognising the importance of a plethora of initiatives that may be available to a firm based on its context, it is argued here that H&S provides excellent evidence of a locus that can be used. For a start, the concepts and principles inherent in lean and sustainability are relevant to the practice of construction management. The relevance of these concepts and principles can be discerned from the number of articles that have been published about them in major construction management journals, and the number of papers that have been presented about them at construction conferences – see the journal *Construction Management and Economics*, for example. It can thus be argued that these concepts will form key aspects of change drivers in the construction industry, and they will play an important role in the realisation of performance criteria in the form of cost, quality, environment, clients, end users, worker satisfaction and H&S, either through the agency of exchange, or through experience. Although the literature has failed to provide sufficient findings that document the interface between H&S and lean and sustainability, important insights were highlighted through various case studies.

The authors of an article based on industrialised housing observed that specific lean strategies appear to have some positive effects on the rate of H&S incidents, which suggests that lean may be beneficial for process improvement, waste reduction, and enhanced H&S in the construction industry. The observations of Court *et al.* (2009) show that lean can reduce H&S-related injuries and accidents. A comparison between the lean construction system in a case project and the traditional method shows that construction workers were exposed to fewer H&S risks from site operations, a situation that resulted in zero reportable accidents (Court *et al.*, 2009). Even on the case project, appropriate ergonomics was achieved through a focus on workplace designs that enhance the well-being of workers. With regard to the use of lean construction in realising sustainability goals, Peng and Pheng (2011) used a case study in Singapore to explore the contribution of the lean concept to the achievement of sustainability objectives in a precast concrete factory. They suggest that the lean production philosophy has practical contributions for sustainable development, which can be adopted by the industry to achieve improved performance in some sustainability factors, including energy consumption, carbon emissions and production efficiency.

To support the discourse, a content analysis was undertaken of IGLC conference proceedings. The number of IGLC H&S and sustainability-related conference papers presented per year from 1998 to 2014 is indicated in Table 6.3. These papers are already lean construction-related, as they have been presented in lean construction annual conferences. It is not possible to indicate the number of H&S and sustainability-related papers that were presented at conferences from 1993 to 1997, since the papers were categorised either under 'theory' or 'safety, quality, and the environment'. The total number of H&S-related papers and sustainability-related papers are equal, at 32 each (see Table 6.3). The sustainability-related papers mostly address the environmental aspects of the traditional model. No paper explicitly addressed the social and economic aspects of sustainability. The H&S-related papers focused on how lean principles can be used to achieve zero accidents in construction.

A brief description of the papers will show that, indeed, lean construction has addressed, and will continue to address, H&S management and sustainability initiatives. For example, Brioso (2011) argues that the integration of loss control, a key feature in H&S management, and lean construction strategies could contribute to a decline in wastes in construction. Another empirical study reinforced the synergy between lean and H&S initiatives (Leino & Elfving, 2011). The findings of the study showed that value exists between lean and H&S through, among other things, respecting people, zero waste and prevention policies. These IGLC papers have evaluated the interface between H&S and lean, and it can be concluded from them that the industry would be better off if it had a synergic approach to these issues.

Table 6.3 IGLC H&S and sustainability-related conference papers

Year	Safety	Sustainability
1998	0	1
1999	0	0
2000	0	0
2001	0	0
2002	2	1
2003	2	1
2004	2	0
2005	2	1
2006	2	0
2007	3	4
2008	1	2
2009	1	2
2010	2	0
2011	4	1
2012	5	7
2013	2	8
2014	4	4
Total	32	32

Similarly, the interface between lean construction and sustainability has been examined to a significant extent in IGLC conferences, particularly at the 2013 conference. A study explored the synergy between lean construction and sustainability in the context of value. Using exemplary lean projects as a unit of analysis, Novak (2012) contends that a strong correlation exists between the cohesiveness of lean thinking and the level of collaboration in terms of delivery of sustainability values. The significance of the study by Novak (2012) is the opportunity for the construct of value to serve as a catalyst that will shift construction management away from restrictive overtones towards a paradigm of positive sustainable prosperity. The study findings show a relation between lean and sustainability, mainly because the stakeholders focused on the concept of value.

Using this focus on value in the examined lean construction papers and the existing management literature as evidence, it can be argued that the proposition depicted in Figure 6.1 warrants further investigation in construction management research. The discussion in the chapter so far has shown that although H&S, lean and sustainability have been well researched, the synergy between them has not been well explored. The synergy between these phenomena is a potential locus where user value can be created at the project level, and shared value can be promoted at the strategic level. The synergy is enabled by the use of appropriate lean construction principles and tools, for the management of both the business and the project aspects of construction. The use of lean principles and tools would also have to consider the economic, environmental and social impact of the work to be done, either on- or off-site.

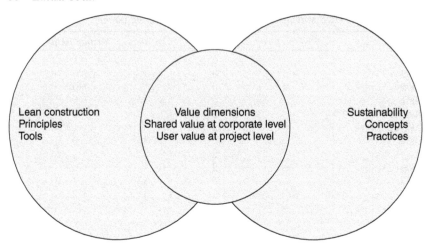

Figure 6.1 Dimensions of value derivable from integrating lean and sustainability

For this reason, this concept of value recognises the claim that there is a need to shift away from management by objectives, so that new methodologies that can advance the performance of projects can be stimulated (Ballard & Howell, 2004). In particular, the thinking behind Figure 6.1 supports the view that accidents and injuries in construction constitute waste (Forbes & Ahmed, 2011). The notion depicted in the figure is congruent with the claims of Ofori (1992) in terms of the potential of lean construction activities to improve H&S, due to their limited environmental impact, and thus to lead to value (Smallwood & Haupt, 2005). Appropriate evaluation of construction and the environment leads to optimal response to H&S issues such as pollution, with their negative impact on the environment (Coble & Kibert, 1994). Dust from construction activities, hazardous materials, and the release of non-biodegradable material into the environment have H&S implications for the general public. Even on a global scale, H&S has to be promoted in a green economy, which is a vehicle for accelerating sustainable development (International Labour Organization, 2012).

The main lean principle is waste, which must be addressed through multiple levels of engagement. The efforts expended on the identification and elimination of wastes impacts on the creation of value at the strategic level and the project level in construction. Creating shared value, as shown in Table 6.4, would emerge from the considerations and deliberations that inform decisions and actions at the interface between lean and sustainability. A survey conducted among construction professionals in the United Kingdom indicates that there are a range of gains associated with integrated implementation of lean construction and sustainability (Ogunbiyi *et al.*, 2014). According to Ogunbiyi *et al.* (2014), benefits of such a synergy include improved corporate image and sustainable competitive advantage, improved process flow and productivity, enhanced environmental quality and greater compliance with client requirements. Ogunbiyi *et al.* (2014) further observe that just-in-time, visualisation tools, value analysis, daily huddle meetings and value stream mapping are the common tools that enable the achievement of sustainability goals. Most importantly, the survey also highlights several areas of linkage between lean and sustainability.

Notable among the areas are waste reduction, value maximisation and H&S. In the study, the connection between H&S, waste and value as interdependent phenomena was established. Similarly, Carneiro *et al.* (2012) observe that the complementarity of lean and sustainability results in the general elimination of waste, and the adding of value for customers. While sustainability is driven by legislation and business needs, strategic options are the main reason for the adoption of lean construction, as is the case in Sri Lanka (Senaratne & Wijesiri, 2008). Because workers are often ignorant of the flow of activities that create waste and their causes, core principles in lean construction and sustainability, in addition to H&S management, can be used to stimulate continual improvement in the sector.

The abstraction suggested in Figure 6.1, however, requires a plan of action in order to operationalise it. The future empirical study which will follow this initial literature review should be able to show how a focus on H&S will improve project value and promote sustainable development. The significant gap in the literature that should be filled will relate to a methodology will allow integration of H&S, lean, and sustainability for delivery of project value in construction. The idea of user value and shared value is a main contribution of this chapter, since within the IGLC community the concept of value (1) has been broadly influenced by the production view of construction – the TFV perspective; (2) is regarded as an ambiguous concept because of different interpretations of the concept, which contributes to an understanding of the concept that includes subjective aspects; (3) refers to the delivery of value at the project level, where waste reduction, planning and control of construction activities are linked to value; and (4) often refers to the fulfilment of customer requirements (Salvatierra-Garrido *et al.*, 2012). The introduction of shared value provides a platform for taking the conceptualisation of value from the project level to the corporate level within the lean research community.

Table 6.4 Considerations related to creating user value and shared value in construction

Lean principles	Sustainability concepts	Creating shared value
Eliminate waste/reduce the share of waste	Economy – individual firm, and market	Value – economic and social benefits in terms of cost
Clearly specify value from the perspective of the ultimate client/increase output value through systematic consideration of customer requirements	Environment – proximity	Joint company and community value creation
Clearly identify the process that delivers what the client values (the value stream), and eliminate all wasteful steps/reduce variability	Society – location and orientation	Integral to competing for business
Reduce the cycle time – to reduce management efforts and interruptions, and to increase delivery time to the customer	An organisation/firm/institution	Integral to profit maximisation
Make the remaining value-adding steps flow without interruption, by managing the interfaces between different steps/simplify by minimising the number of steps, parts and linkages	People within an organisation – in a corporate context	Agenda is company-specific and internally generated

Realigns the entire company budget

Let the client pull – do not make anything until it is needed, and when it is needed, make it quickly/increase output flexibility

Increase process transparency – use visual controls plus implementation of information systems to reduce the tendency for errors in the process

Focus control on the complete process – evaluate and control the process as a whole, as opposed to segmental control of the process

Pursue perfection by continual improvement/build continual improvement into the process

Balance flow improvement with conversion improvement

Benchmark – know the process, assess the strengths and weaknesses of the sub-process; know the industry leaders and find, understand and then best practices

Source: developed from Porter & Kramer (2011); Koskela (1992); Constructing Excellence (2004); WCED (1987)

6.5 Summary

Given the documented need to change the *pace and functionality* of delivered projects, a paradigm shift in relation to the conceptualisation of value in construction is crucial in the sector. Lean construction and sustainability have emerged as key drivers of such a shift. Through the established complementarity of lean and sustainability, waste in the construction process can be addressed at both the strategic level and the project level. The conceptual idea in this chapter argues for exploitation of the link between lean and sustainability, in order to create two main dimensions of value: user value and shared value. Shared value, which is promoted and created at the corporate level of an enterprise, subsumes user value, which is often created at the project level. Within the IGLC community, value has been conceptualised primarily at the project level. This chapter, however, introduces the possibility of value creation at the corporate level, using the shared value perspective. It is notable that the three avenues for creating shared value which were indicated in the chapter are mutually reinforcing, and their use would benefit construction firms and the communities in which the firms operate. An example of how shared value can be created is transformation of the procurement system, so that quality and productivity can be increased. There are various ways in which addressing social concerns can yield productive gains to a contractor. For example, consider what happens when a contractor invests in a wellness programme for their stakeholders. Society benefits, as employees of the firm, as well as their families, become healthier, and low morale and absenteeism are minimised. The integrated project delivery (IPD) mechanism serves as a point of reference for procurement. However, this concept is limited, in that it is yet to be empirically investigated. The application of the concept depicted in Figure 6.1 to a case project should provide insights into how dimensions of value will emerge from the synergy between lean and sustainability.

References

Ballard, G. & Howell, G.A. (2004), Competing construction management paradigms. *Lean Construction Journal*, Vol. 2004, pp. 38–45.

Boztepe, S. (2007), User value: Competing theories and models. *International Journal of Design*, Vol. 1(2), pp. 55–63.

Brioso, X. (2011), Applying lean construction to loss control. In: *Proceedings of the 19th Conference of the International Group for Lean Construction*, Lima, Peru, 13–15 July, pp. 656–666.

Carneiro, S.B.M., Campos, I.B., De Oliveira, D.M. & Barros Neto, J.P. (2012), Lean and green: A relationship matrix. In: *Proceedings of the 20th Conference of the International Group for Lean Construction*, San Diego, CA, 18–20 July, pp. 61–70.

Carruthers, B. & Babb, S. (1996), The color of money and the nature of value: Greenbacks and gold in postbellum America. *American Journal of Sociology*, Vol. 101(6), p. 1556.

Coble, R.J. & Kibert, C.J. (1994), The environment as a construction safety concern. In: *Proceedings of the 5th Annual Rinker International Conference Focusing on Construction Safety and Loss Control*, Gainesville, FL, 12–14 October, pp. 535–542.

Constructing Excellence (2004), *Lean Construction*, Constructing Excellence, London.

Court, P.F., Pasquire, C. & Gibb, A. (2009), A lean and agile construction system as a set of countermeasures to improve health, safety and productivity in mechanical and electrical construction. *Lean Construction Journal*, Vol. 2009, pp. 61–76.

De Valence, G. & Runeson, G. (2011), On the state of the building industry after the GFC and the Euro crisis. *Australasian Journal of Construction Economics and Building*, Vol. 11(4), pp. 102–113.

Forbes, L.H. & Ahmed, S.M. (2011), *Modern Construction: Lean Project Delivery and Integrated Practices*, CRC Press, Boca Raton, FL.

Ganda, F. & Ngwakwe, C.C. (2014), The role of social policy in transition towards a green economy: The case of South Africa. *Environmental Economics*, Vol. 5(3), pp. 32–41.

Graeber, D. (2001), *Toward an Anthropological Theory of Value: The False Coin of Our Own Dreams*, Palgrave, New York.

Green, S.D. (2011), *Making Sense of Construction Improvement*, Wiley-Blackwell, Chichester.

International Labour Organization (ILO) (2012), *Promoting Safety and Health in a Green Economy*, ILO, Geneva.

Kamakura, W.A. & Novak, T.P. (1992), Value-system segmentation: Exploring the meaning of LOV. *Journal of Consumer Research*, Vol. 19(1), pp. 119–132.

Kibert, C.J. (2009), *Sustainable Construction: Green Building Design and Delivery*, 2nd edition, Wiley, Hoboken, NJ.

Koskela, L. (1992), Application of the new production philosophy to construction. Technical Report No. 72, Center for Integrated Facility Engineering (CIFE), Stanford University.

Koskela, L., Bolviken, T. & Rooke, J. (2013), Which are the wastes of construction? In: *Proceedings of the 21st Annual Conference of the International Group for Lean Construction*, Fortaleza, Brazil, 29 July–2 August.

Leino, A. & Elfving, J. (2011), Last planner and zero accidents program integration: Workforce involvement perspective. In: *Proceedings of the 19th Conference of the International Group for Lean Construction*, Lima, Peru, 13–15 July, pp. 622–632.

Ma, U. (2011), *No Waste: Managing Sustainability in Construction*, Gower, Surrey.

Maltz, E. & Schein, S. (2013) Cultivating shared value initiatives. *Journal of Corporate Citizenship*, Vol. 47, pp. 55–74

Matar, M.M., Georgy, M.E. & Ibrahim, M.E. (2008), Sustainable construction management: Introduction of the operational content space (OCS). *Construction Management and Economics*, Vol. 26(2), pp. 261–275.

Mollenkopt, D., Stolze, H., Tate, W.L. & Ueltschy, M. (2010), Green, lean, and global supply chains. *International Journal of Physical Distribution & Logistics Management*, Vol. 40(1/2), pp. 14–41.

Novak, V.M. (2012), Value paradigm: Revealing synergy between lean and sustainability, In: *Proceedings of the 20th Conference of the International Group for Lean Construction*, San Diego, CA, 18–20 July, pp. 51–60.

Ofori, G. (1992), The environment: The fourth construction project objective? *Construction Management and Economics*, Vol. 10(5), pp. 369–395.

Ogunbiyi, O., Oladapo, A. & Goulding, J. (2014), An empirical study of the impact of lean construction techniques on sustainable construction in the UK. *Construction Innovation*, Vol. 14(1), pp. 88–107.

Park, J. & Han, S.H. (2013), Defining user value: A case study of a smartphone. *International Journal of Industrial Ergonomics*, Vol. 43(5), pp. 274–282.

Porter, M.E. & Kramer, M.R. (2006), Strategy and society: The link between competitive advantage and corporate social responsibility. *Harvard Business Review*, Vol. 84(12), pp. 78–92.

Porter, M.E. & Kramer, M.R. (2011), Creating shared value. *Harvard Business Review*, Vol. 89(1/2), pp. 62–77.

Salvatierra-Garrido, J., Pasquire, C. & Miron, L. (2012), Exploring value concept through the IGLC community: Nineteen years of experience, In: *Proceedings of the 20th Conference of the International Group for Lean Construction*, San Diego, CA, 18–20 July, pp. 361–370.

Senaratne, S. & Wijesiri, D. (2008), Lean construction as a strategic option: Testing its suitability and acceptability in Sri Lanka. *Lean Construction Journal*, Vol. 2008, pp. 34–48.

Shen, L.Y., Song, S.C., Hao, J.L. & Tam, V.W.Y. (2008), Collaboration among project participants towards sustainable construction: A Hong Kong study. *The Open Construction & Building Technology Journal*, Vol. 2(1), pp. 59–68.

Smallwood, J. & Haupt, T. (2005), The need for construction health and safety (H&S) and the construction regulations: Engineers' perceptions. *Journal of the South African Institution of Civil Engineering*, Vol. 47(2), pp. 2–8.

Warnock, A.C. (2007), An overview of integrating instruments to achieve sustainable construction and building. *Management of Environmental Quality: An International Journal*, Vol. 18(4), pp. 427–441.

WCED (1987), *Our Common Future: Report of the World Commission on Environment and Development*, United Nations General Assembly, New York.

Womack, J.P. & Jones, D.T. (1994), From lean production to the lean enterprise. *Harvard Business Review*, Vol. 72(2), pp. 93–103.

Wu, Y.C. (2003), Lean manufacturing: A perspective of lean suppliers, *International Journal of Operations & Production Management*, Vol. 23(11), pp. 1349–1376.

Part III

Control of waste in construction

7 Last Planner System

Improving planning procedures to reduce waste

Søren Lindhard

In this chapter the focus is on the Last Planner System (LPS) of production control and how it contributes to the reduction of waste. LPS is a central tool in the lean construction toolbox and is based on the lean construction philosophy and its three-part focus on transformation, flow and value. In accordance with the lean construction philosophy, removing waste will directly improve the transformations and flows and indirectly improve the value creation by allowing more value to be added to the end product. LPS is focused on improving the flows to optimise the production process and increase the production output. The chapter offers insights into how elements in LPS contribute to the reduction of waste.

7.1 Background

Lean construction is based on the transformation–flow–value (TFV) theory introduced by Koskela (1992). The traditional transformational view is expanded to also include the undergoing, moving, waiting, and inspection activities, and thus includes the sub-processes necessary to complete the construction project. Only transformations create value, thus the additional processes are regarded as waste (Lindhard & Wandahl, 2012a). In a lean perspective the transformations should be trimmed to be as value-creating as possible, while the non-value adding activities should be reduced or removed, if possible (Koskela, 1992, 1996). It is important to be aware that some of the waste activities are necessary in order to carry-out the value-creating activities. For instance, transportation of material from factory to site could be regarded as necessary waste, because it is necessary in order to initiate the downstream activities. In lean construction, there is a focus on creating value for the customer. Value is understood as the fulfilment of the demands and requirements stated by the end customer.

In brief, in lean production, the three-part focus on TFV focuses on producing outputs as efficiently as possible, eliminating waste processes and increasing value creation. The focus in this book is on the waste perspective; therefore this chapter will focus on how waste is handled. According to the lean philosophy, waste can be categorised into *muda* (waste of resources), *muri* (waste by overburdening) and *mura* (waste by variations). In relation to the TFV theory, waste can be removed by:

- reducing lead time;
- reducing variation;
- increasing process transparency;
- simplifying the production; and
- hampering sub-optimisation (Koskela, 1992, 2000).

7.1.2 Last Planner System of production control

The LPS of production control was introduced by Ballard (2000) to increase the schedule quality and reliability. The motivation for developing LPS was a research study conducted by Ballard and Howell (1994). The research looks into schedule reliability and found that only about half of work tasks were completed on schedule. After applying LPS, the schedule reliability was successfully increased. Approximately 70 per cent of work tasks were completed on schedule (Lindhard & Wandahl, 2014a). Moreover, other positive effects have been reported, such as increased labour productivity and direct work (Ballard, 1999).

This chapter includes a short introduction to LPS. An in-depth description can be found in Ballard (2000) and amendments can be found in Ballard and Howell (2003a). LPS consists of four overall schedules, namely the master schedule, the phase schedule, the look-ahead schedule, and the weekly work plans, as highlighted below.

The master schedule. The master schedule is the main schedule and contains the deadlines and milestones for the entire construction process. The lower-level schedules focus on ensuring that the deadlines in the master schedule are adhered to.

The phase schedule. The phase schedule focuses on identifying in what sequence the activities should be completed. At the outset the deadlines and milestones are defined in the master schedule. The sequence is determined by working backwards from the deadlines and identifying activities and handoffs between the work crews (Hamzeh *et al.*, 2008). In this process it is important to determine the duration of the activities and the interdependencies and relationship between the activities. To ensure optimisation of the phase schedule, every contractor involved in the construction project should participate in determining the sequence by providing inputs based on his or her insights, expertise and specific specialisation (Howell, 1999).

The look-ahead schedule. The look-ahead schedule focuses on make-ready activities, thereby ensuring that the activities can be completed. By removing all obstacles, the look-ahead schedule ensures that it is possible to complete the activities in the determined sequence, according to the phase schedule. The determination of an activity's readiness is based on the fulfilment of the seven preconditions introduced by Koskela (1999). Thus, if just one of the seven preconditions is not fulfilled, the activity cannot be completed. The seven preconditions are as follows:

1   Design: the correct plans, drawings and specifications need to be available.
2   Materials: the correct materials and components need to be available.
3   Workers: a qualified workforce needs to be available.
4   Equipment: the correct equipment and machinery need to be available.
5   Space: There needs to be enough work space to allow the work task to be carried out.
6   Connecting works: previous activities need to have been completed.
7   External preconditions: all external factors need to be in order.

Lindhard and Wandahl (2012b) later divided the external category into safety (there needs to be safe working conditions), climate conditions (climate conditions must be under control and acceptable) and know working conditions (the existing and surrounding conditions which affect the work task need to be known). When an activity is free from constraints it is moved to a buffer, and in this way forms part of a workable backlog. When drawing up the weekly work plans, only activities from the workable backlog are included, thereby ensuring that only ready activities are included in the work schedules. It is important to be aware of possible constraints which may arise and which may affect the readiness of the activities. Thus, any event affecting the readiness of the activities must be registered and the backlog updated. To ensure that there are enough activities ready for work, it is suggested to keep the backlog of activities on two weeks of work (Ballard, 1997).

The weekly work plans. The weekly work plans contain the work which has to be carried out on the construction site during the following week. The weekly work plans detail the work tasks that should be completed within the work week and when they must be done. When selecting work tasks, only work tasks from the backlog of ready work are selected. This ensures that all selected work tasks can be carried out. A central aspect of LPS is to follow-up on the work and to adhere to the defined schedule. As a measurement tool to schedule quality, the percentage planned completed (PPC) measurement was introduced. The PPC measurement is a simple comparison between the planned completed and the actual completed. In this way the activities which not were completed according to the schedule are identified (Ballard, 2000). The PPC measurement provides feedback to the site manager, who can easily determine whether he needs to intervene. Both activities which are completed too early and those completed too late are identified and the underlying reason for this is determined by applying the Five Whys. The Five Whys is a process whereby one will continue to ask 'Why?' until the underlying cause is revealed. The root cause analysis forms the learning part of LPS: by learning from mistakes, repetitions can be avoided and the schedule quality can be improved (Lindhard & Wandahl, 2014b).

7.2 Last Planner System and waste

Lean construction is based on the TFV theory, but LPS focuses only on improving the flow. Therefore, LPS does not improve the actual transformation or increase the value creation. In relation to waste, LPS has elements which reduce *muda, mura* and *muri*, but the focus is on improving the flow of work. As stated by Schonberger (1986), 'Variation is the universal enemy'. Variation is everywhere and exists in many forms in on-site construction (Thomas *et al.*, 2003). It affects both the product and the production workflow (González *et al.*, 2010):

- Product variation affects the output quality and results in increased rework (Petersen, 1999).
- Production workflow variation affects the work processes and decreases productivity (González *et al.*, 2010).

LPS focuses on decreasing workflow variation, gaining control of the production output and thereby increasing productivity. Variation in the production throughput results in activities being completed either before or after the scheduled deadline. Positive variation refers to finishing a work task before the deadline, while negative variation refers to finishing a work task after the deadline (Lindhard, 2014). While negative variation (finishing a work task after the deadline) causes delay and disruptions to the schedules (Ballard & Howell, 1994), positive variation (finishing a work task before the deadline) creates gaps in production and results in unexploited capacity (Yeo & Ning, 2006). Therefore both are undesirable in a production system.

LPS was introduced to increase schedule reliability (Howell & Ballard, 1995). Increased schedule reliability will implicitly lead to reduced variation and increased on-site productivity. Thus, because the primary focus in LPS is to increase schedule reliability, it is necessary that the schedule must be protected from variation occurring anywhere on-site. LPS has two different approaches in relation to reducing variation and thus to stabilising the production workflow:

1 Shielding the production against the causes which trigger variation.
2 Having buffers with ready work tasks to absorb variations which slip past the shield.

An in-depth description of how LPS handles waste caused by variation (*mura*) is presented in Section 7.3. In LPS, the stabilisation of the production workflow is the main approach for the removal of waste. While a stabilised production workflow directly removes waste of variation (*mura*), it indirectly removes both waste of overburdening (*muri*) and waste of resources (*muda*). In addition to the stabilisation of the production workflow, LPS has several elements which also affect the waste caused by *muri* and *muda*.

In relation to *muda*:

- The phase schedule and the associated sequencing process help to optimise the order in which the work tasks are completed.
- The just-in-time principle applied to the ordering of materials, also called the pull principle, helps to reduce the storage needs of materials.

In relation to *muri*:

- The matching of work tasks in the weekly work plans to the present work capacity ensures that the workforce is not overburdened.

In-depth descriptions of how LPS handles waste caused by resources (*muda*) and overburdening (*muri*) are presented in Sections 7.4 and 7.5. LPS strives for continuous improvement by learning from previous mistakes. Improving by learning from previous mistakes to avoid repetition enables the reduction of *muri*, *muda* and *mura*. An introduction to how learning is facilitated in LPS is presented in Section 7.6.

7.3 Decreasing variation (*mura*)

LPS reduces waste by stabilising the workflow. A stabilised workflow makes it easier to match resources and capacity, and results in decreased variation (*mura*) and increased productivity (Tommelein *et al.*, 1999). Lindhard (2014) found that the waste caused by variation in the production workflow results between handoffs. A stabilised workflow is achieved by reducing inflow variation and by keeping a backlog of ready work to absorb incoming variation. Inflow variation is reduced by applying the making-ready process. Thus, by ensuring that every precondition is fulfilled, and by only selecting activities from the weekly work plans which are ready, one ensures that the scheduled activities can actually be completed. By ensuring that every activity in the schedule can be completed, the number of interruptions in the production flow is reduced (Ballard & Howell, 1994) and therefore the likelihood of completing the scheduled work task on time and thus the likelihood of keeping to the schedule increases.

In the weekly work plans, the work tasks which are ready are selected and matched to the present work capacity. The subcontractors are part of the weekly scheduling, which, owing to their practical and professional knowledge and insights, increases the schedule quality. Moreover, by mutually agreeing and committing to the schedule, schedule observance is increased. According to Ballard (1994), honesty is also increased, thus 'We do what we say we are going to do' because we know what is possible and which activities which can actually be completed. Thus, it helps to move beyond the can-do mentality where even poorly prepared and unprepared work tasks are accepted (Howell & Ballard, 1997). The backlog of ready work serves as a buffer that absorbs the variation which gets through the stabilised and shielded workflow. Non-ready work activities can quickly be replaced by ready work from the backlog, thus production can be kept on schedule (Ballard, 1994; Howell & Ballard, 1994). Besides the backlog of ready work, buffers between work tasks can be used to decrease the interdependencies between the work tasks and to absorb variation in work pace of upstream work tasks without creating interruptions or affecting the downstream work activities, thus maintaining the production flow.

Even though buffers can help stabilise the production flow, they need to be applied wisely because buffers are costly. The associated cost of buffering includes an increased idle inventory and usage of storage space, buffer fill time, double handling of materials, and inventory management, among others (Lindhard & Wandahl, 2014b). In a perfect world, buffers should be removed because buffers are waste. Therefore, buffers are an expensive approach to reducing variation and should be used carefully. When using buffers, the size of the buffer should be minimised to fit actual requirements derived from the likelihood and possibility of variation in upstream activities (Howell & Ballard, 1994). The introduction of buffers underlines LPS focus on variation, where the importance of securing a stabilised workflow outweighs the waste introduced by applying buffers. In other words, buffers are considered necessary waste to absorb incoming variation and thus an approach to reduce variation in the production workflow to increase on-site productivity.

7.4 Decreasing waste of resources (*muda*)

Even though LPS does not focus directly on reducing the waste of resources because it does not consider how the transformations are completed, it still includes factors which reduce *muda*. One central element in LPS is the sequencing of activities. During the sequencing, which is a part of the phase scheduling process, the best possible sequence of the work tasks is selected. With the setting out of deadlines in the master schedule the work tasks are organised according to their duration and their interdependencies. Planning and optimising the sequence of work tasks is an essential part of LPS and has a major effect on the quality and reliability of the schedule. By improving the sequence, unnecessary activities can be removed (Ballard, 2000), which reduces a number of wastes such as reduced transportation and movement on-site, reduced re-handling of materials and reduced waiting and idle time caused by interrelated activities. Moreover, the phase scheduling process helps to break down activities into smaller batches. By reducing the batch size, the reliability is increased, thus all project participants have an increased commitment towards the upcoming work tasks (Ballard & Howell, 2003b).

To gain control of the upcoming construction project, a major problem is determining the interdependencies and handoffs across work scopes. Owing to a general increase in construction complexity, revealing the underlying interdependencies between work tasks and work crews has become more difficult and important in relation to achieving control of the production process (Jang & Kim, 2008). During the phase scheduling process all subcontractors need to be present and take part in determining the schedule because it increases the schedule quality. The reason is that many interdependencies are only discovered when the subcontractors start to plan work tasks across work scopes (Ballard & Howell, 2003a). In LPS, materials are ordered by having a pull approach. Thus, instead of determining a set of fixed delivery dates, ordering of materials is handled much more flexibly by applying the just-in-time principle. The difference is that materials are now ordered and delivered to the construction site only when needed. The effect is decreased storage and re-handling of materials. Moreover, by reducing the storage of materials on-site, the likelihood and cost of loss of materials either by theft or damage decreases. Loss of materials is a waste of resources. Therefore the pull principle reduces the waste of *muda*.

The stabilisation of the workflow also reduces waste of resources. As mentioned earlier, a stable workflow makes it easier to match work tasks to capacity, to select the right amount of work and to secure a high utilisation of the capacity of the present labour and machinery. By increasing utilisation rates, the full potential of the production system is harnessed, resulting in increased throughput and productivity. Ballard (1999) warns against planning for a 100 per cent utilisation rate and states that it is 'better to under load production units in order to allow for variability in production'. By under loading the capacity and keeping the utilisation just below 100 per cent, the schedule will not be affected by small variations in throughput. Therefore, the likelihood of maintaining the schedule increases, which improves downstream performance and productivity (Ballard, 1999).

7.5 Decreasing the risk of overburdening (*muri*)

Putting excessively high pressure on the capacity of both the manpower and the machinery together with increases in work task complexity creates the risk of overburdening the workers on-site. Overburdening in lean is called *muri*, and is the waste which occurs when the work tasks become confusing, chaotic and overwhelming. If the skills of the workers do not match the complexity of the work task, the workers simply do not have the ability to perform the work task at a satisfactory level. If the amount of work is not matched to the capacity of the workforce, the workers become overloaded by the increased work pressure, they become stressed and they begin to make mistakes (Kiev & Kohn, 1979). Owing to the decreasing likelihood of successfully keeping to the schedule, they begin to lose morale and become pessimistic about the project (McClelland *et al.*, 1976). Therefore, overburdening the workers on-site will result in decreased output quality. Decreased output quality increases the need for rework and thus induces waste.

The following two aspects of LPS reduce the risk of overloading the workers conducting the work tasks on-site:

1   stabilisation of the production workflow;
2   matching of work tasks in the weekly work plans to the present work capacity.

Stabilisation of the production workflow creates a steady workflow. Because variations are reduced either by the inflow shield or by the buffer, the concomitant fluctuations in work pressure caused by unpredictable, complex and varying workflow are reduced. Moreover, a stabilised workflow makes it easier to match the work tasks in the weekly work plans to the present work capacity because the likelihood of keeping to the schedule is increased. A better matching of work tasks to capacity makes it easier to avoid creating excessive work pressure and creating a mismatch between competences and task complexity. Therefore, the result is a reduced risk of overburdening the workers present on-site.

7.6 Continuous removal of waste in future processes

In LPS the primary focus is on improving schedule reliability. LPS also includes an element to continuously improve the schedule reliability and thus to reduce the waste caused by variations in the production workflow. To achieve this improvement, the actual schedule reliability needs to be determined and analysed. Therefore, to distinguish between quality failures and schedule failures (failures to execute conducted plans) the PPC measurement was introduced (Ballard, 1994; Ballard & Howell, 1994).

After a completed work week, the completed activities are compared to the scheduled work tasks and the PPC measurement is calculated to determine the adherence to the determined schedule and thus the schedule quality. During the calculation of the PPC measurement, the work tasks not completed are registered. Afterwards root causes are investigated by applying the Five Whys. This is done to identify the triggers and to avoid future repetitions.

The root-cause identification constitutes the learning element in LPS. Ballard (1994) states that 'the starting point for improvement in planning is measuring the percentage of planned activities completed, identifying reasons for non-completion, and tracing reasons back to root causes that can be eliminated to prevent repetitions'. Moreover, the PPC measurement increases the commitment to learning and to continuously improve. The result is a gradually improved production process where the risk of variations and interruptions is reduced, which reduces waste by increasing the likelihood of adhering to the determined schedule.

7.7 Summary

In this chapter the focus has been on the Last Planner System (LPS) approach to handling and reducing waste. LPS is a lean construction tool that is based on the transformation–flow–value theory, which focuses on improving the workflow. LPS has several built-in elements which help to reduce waste. In this chapter these elements have been divided into factors which reduce *muda* (waste of resources), *muri* (waste of overburdening) and *mura* (waste of variation).

LPS was developed to increase schedule reliability. Schedule reliability in LPS is achieved by stabilising the production workflow. Therefore the focus in LPS is on reducing variation in the production workflow, also called *mura*. To achieve increased schedule reliability, LPS protects the production flow by ensuring that the triggers to variation are reduced. The triggers are removed during the making-ready process, where every one of the seven preconditions is fulfilled. Fulfilling the seven preconditions ensures that the work tasks are ready to be completed and that the production flow can therefore continue without interruption. Moreover, by buffering ready work tasks, the variation which slips through the making-ready process is absorbed with only minimal effect in the production workflow.

A stabilised workflow indirectly has a positive effect on *muri* and *muda*. Besides the stabilised workflow, the waste of *muda* is reduced by applying the phase schedule, where interdependencies and handoffs between work crews are identified and, based on these, the sequence is optimised. While optimising the sequence, waste in the form of unnecessary activities is removed. Moreover, the application of the just-in-time principle, where materials are drawn only when needed, reduces the storage and loss of materials. In the weekly work plans the waste of *muri* is reduced by matching the selected work task to the present work capacity. Selecting ready work tasks only from the ready buffer makes the matching process easier. The result is the likelihood of complying with the weekly work plans and thus a less stressed or overburdened workforce.

Finally, LPS has a built-in learning element to ensure a learning process. By measuring the PPC, interruptions in the workflow are identified. Afterwards the Five Whys can be applied to identify the root cause, which is helpful in an attempt to avoid future repetitions and thus help to reduce the waste of *muri*, *muda* and *mura*.

References

Ballard, G. (1994). Implementing lean construction: Stabilising workflow. In: *Proceedings for the 2nd Annual Conference of the International Group for Lean Construction*, Santiago, Chile, 28–30 September, pp. 101–110.

Ballard, G. (1997). Look ahead planning: The missing link in production control. In: *Proceedings for the 5th Annual Conference of the International Group for Lean Construction*, Gold Coast, Australia, 16–17 July, pp. 13–26.

Ballard, G. (1999). Improving workflow reliability. In: *Proceedings for the 7th Annual Conference of the International Group for Lean Construction*, Berkeley, CA, 26–28 July, pp. 275–286.

Ballard, G. (2000). The Last Planner System of production control. PhD thesis, University of Birmingham.

Ballard, G. & Howell, G. (1994). Implementing lean construction: Improving downstream performance. In: *Proceedings for the 2nd Annual Conference of the International Group for Lean Construction*, Santiago, Chile, 28–30 September, pp. 111–125.

Ballard, G. & Howell, G. (2003a). An update on last planner. In: *Proceedings for the 11th Annual Conference of the International Group for Lean Construction*, Blacksburg, VA, 22–24 July.

Ballard, G. & Howell, G. (2003b). Lean project management. *Building Research & Information*, Vol. 31(2), pp. 119–133.

González, V., Alarcón, L.F., Maturana, S., Mundaca, F. & Bustamante, J. (2010). Improving planning reliability and project performance using the reliable commitment model. *Journal of Construction Engineering and Management*, Vol. 136(10), pp. 1129–1139.

Hamzeh, F.R., Ballard, G. & Tommelein, I.D. (2008). Improving construction workflow: The connective role of lookahead planning. In: *Proceedings for the 16th Annual Conference of the International Group for Lean Construction*, Manchester, UK, 16–18 July, pp. 635–644.

Howell, G. (1999). What is lean construction – 1999? In: *Proceedings for the 8th Annual Conference of the International Group for Lean Construction*, Berkeley, CA, 26–28 July, pp. 1–10.

Howell, G. & Ballard, G. (1994). Implementing lean construction: Reducing inflow variation. In: *Proceedings for the 2nd Annual Conference of the International Group for Lean Construction*, Santiago, Chile, 28–30 September.

Howell, G. & Ballard, G. (1995). Factors affecting project success in the piping function. In: *Proceedings for the 3rd Annual Conference of the International Group for Lean Construction*, Alberquerque, NM, 16–19 October.

Howell, G. & Ballard, G. (1997). Lean production theory: Moving beyond 'can do'. In: *Proceedings for the 2nd Annual Conference of the International Group for Lean Construction*, Santiago, Chile, 28–30 September.

Jang, J.W. & Kim, Y.W. (2008). The relationship between the make-ready process and project schedule performance. In: *Proceedings for the 16th Annual Conference of the International Group for Lean Construction*, Manchester, UK, 16–18 July, pp. 647–656.

Kiev, A. & Kohn, V. (1979). *Executive Stress*, American Management Association, New York.

Koskela, L. (1992). Application of the new production philosophy to construction. PhD thesis, Stanford University.

Koskela, L. (1996). Towards the theory of (lean) construction. In: *Proceedings for the 4th Annual Conference of the International Group for Lean Construction*, Birmingham, UK, 26–27 August.

Koskela, L. (1999). Management of production in construction: A theoretical view *Proceedings for the 8th Annual Conference of the International Group for Lean Construction*, Berkeley, CA, 26–28 July, pp. 241–252.

Koskela, L. (2000). *An Exploration Towards a Production Theory and its Application to Construction*, VTT Building Technology, Espoo.

Lindhard, S. (2014). Understanding the effect of variation in a production system. *Journal of Construction Engineering and Management*, Vol. 140(11), pp. 1–8.

Lindhard, S. & Wandahl, S. (2012a). Adding production value through application of value based scheduling, *COBRA 2012: RICS International Research Conference*, Las Vegas, NV.

Lindhard, S. & Wandahl, S. (2012b). Improving the making ready process: Exploring the preconditions to work tasks in construction. In: *Proceedings for the 20th Annual Conference of the International Group for Lean Construction*, San Diego, CA, 18–20 July.

Lindhard, S. & Wandahl, S. (2014a). Exploration of the reasons for delays in construction. *International Journal of Construction Management*, Vol. 14(1), pp. 47–57.

Lindhard, S. & Wandahl, S. (2014b). Scheduling of large, complex, and constrained construction projects: An exploration of LPS application. *International Journal of Project Organisation and Management*, Vol. 6(3), pp. 47–57.

McClelland, D.C., Atkinson, J.W., Clark, R.A. & Lowell, E.L. (1976). *The Achievement Motive*, Irvington, Oxford.

Petersen, P.B. (1999). Total quality management and the Deming approach to quality management. *Journal of Management History*, Vol. 5(8), pp. 468–488.

Schonberger, R.J. (1986). *World Class Manufacturing: The Lessons of Simplicity Applied*, The Free Press, New York.

Thomas, H.R., Horman, M.J., Minchin, R.E. & Chen, D. (2003). Improving labor flow reliability for better productivity as lean construction principle. *Journal of Construction Engineering and Management*, Vol. 129(3), pp. 251–261.

Tommelein, I.D., Riley, D.R. & Howell, G.A. (1999). Parade game: Impact of workflow variability on trade performance. *Journal of Construction Engineering and Management*, Vol. 125(5), pp. 304–310.

Yeo, K.T. & Ning, J.H. (2006). Managing uncertainty in major equipment procurement in engineering projects. *European Journal of Operational Research*, Vol. 171(1), pp. 123–134.

8 Guidelines and conditions for implementing kanban in construction

Dayana B. Costa and André Perroni de Burgos

Due to difficulties in managing material flow during construction, project managers are currently looking for ways to facilitate communication between managerial teams and workers, as well as to control material consumption and flow. The kanban is a technique that may be used to minimise these problems, reduce different types of waste and keep projects within their budgets. The main objective of this chapter is to present guidelines and conditions for effective implementation of kanban in construction processes, based on four empirical studies conducted on projects in Brazil. These projects were evaluated according to a set of criteria for kanban implementation, including mechanisms to reduce wastes, mechanisms to achieve continuous improvement, tools to improve transparency, just-in-time and pull production. This chapter contributes to the understanding of the necessary conditions for kanban implementation to reduce waste in construction.

8.1 Background

The construction industry has traditionally been one of the largest industries in many countries – for example, the USA and Brazil. However, studies have shown a high level of wasted resources in labour and materials, as well as low productivity (Forbes & Ahmed, 2011). For example, 30 per cent of construction costs are typically due to inefficiencies, mistakes, delays and poor communication. In addition, between 25 per cent and 50 per cent of construction costs are lost due to waste and inefficiencies in labour and material control (Forbes & Ahmed, 2011). Elimination of waste is the most important principle in lean production (Shingo, 1989). According to Ohno (1988), to eliminate waste there is a need to start producing only the things required, using minimal resources. This can be linked to the concept of just-in-time, which means that the supply of each process should consist of only what is required, when it is required and in the quantity required (Lu, 1985). Materials management serves the purpose of providing the right materials when they are required, at an acceptable cost. This involves specifying the required materials, acquiring them from suppliers and distributing them to construction sites (Arbulu *et al.*, 2003). Failing to allow a continuous flow of materials will have a negative effect on labour productivity and project costs, and will also increase waste. Thus, material management is an essential part of the production system for a construction project (Khalfan *et al.*, 2008).

In this context, use of the kanban technique can be an effective method to improve the ways that the construction industry manages material and labour resources, thereby improving productivity, reducing waste and keeping projects within their budgets. The kanban technique is a lean approach that evolved in the context of the automotive industry, and it was originally developed by Toyota Motor Company for the purpose of managing and ensuring the flow and production of materials, and supporting production control and performance. The signaling device works as a mechanism to pull materials and parts through a value stream on a just-in-time basis (Shingo, 1989; Monden, 1983; Ohno, 1988). In Japanese, the word kanban means 'card' or 'sign', and is the name given to the inventory control card used in a pull system (Productivity Press Development Team, 2002). However, like other techniques/tools at Toyota, the kanban technique was created to fulfil the specific need of Toyota to work effectively under specific production and market conditions. Because these conditions are not the same for all organisations, the kanban technique is limited in the form in which it is reported in the literature. For instance, the kanban is not adequate when demand is unstable, processing times are unstable, operations are not standardised, there are long setup times, there is a wide variety of items, or the supply of raw materials is uncertain (Monden, 1983; Ohno, 1988). Due to difficulties in using the kanban in its original form in diverse situations, variations, or adaptations, of the kanban have been created, so as to properly adapt to the specific realities of companies (Junior & Filho, 2010); this has also occurred in the construction industry.

The first studies applying the just-in-time concept and kanban in construction were proposed by Pheng and his research groups (Low & Chan, 1997; Low & Mok, 1999; Low & Choong, 2001). Low and Mok (1999) reported the application of the just-in-time concept for site layout, to improve productivity and quality. Arbulu *et al.* (2003) used supplier kanbans to indicate the need for replenishment of selected products from preferred suppliers to the site, with the goal of simplifying the process of acquiring, storing, distributing and disposing of selected made-to-stock products on-site. Then, Arbulu (2009) described the benefits of using kanbans to manage the supply of a large number of non-task-specific materials in a large airport project. Jang and Kim (2007) explored the use of kanbans as a work order for the construction production process and safety control. Khalfan *et al.* (2008) reported on the use of the kanban to deliver selected products from suppliers and off-site manufacturers on a just-in-time basis. In Brazil, several applications of kanbans for managing material handling and delivery on-site in housing projects have been reported in the literature (see Kemmer *et al.*, 2006; Heineck *et al.*, 2009; Burgos & Costa, 2012; Barbosa *et al.*, 2013; Menezes, 2013). Although some applications have been developed in recent years that attempt to improve material flow, use of this technique is still limited in construction sites. The objectives of this chapter are: (1) to provide an overview of the kanban technique and its principles; (2) to describe and assess the implementation of kanban in four Brazilian building projects, identifying opportunities for improvements in the efficient use of the technique; and (3) to establish the necessary conditions and guidelines for the implementation of kanban in construction processes, with the goal of reducing waste.

8.2 Lean principles behind kanban

Before discussing the kanban technique, it is essential to present the core principles of the Toyota Production System (TPS), which form the basis of kanban. As mentioned, the kanban is a way to manage the just-in-time production method (Monden, 1983), which was developed to adapt to changes caused by troubles in the processes, as well as changes in demand, by having all processes produce the required goods at the required time, in the required quantities, without generating stock (Monden, 1983; Shingo, 1989). For Monden (1983), the first requirement for just-in-time production is to enable all processes to achieve accurate timing and the required quantities. If the just-in-time concept is realised throughout a business, then unnecessary inventories in the factory will be eliminated, thereby making storehouses or warehouses unnecessary (Monden, 1983). Therefore, a company seeking to attain just-in-time functionality is attempting to work systematically, to reduce excess work, to eliminate wastes, and to increase productivity (Lu, 1985).

The kanban technique can be described in 'pull' and 'push' terms; material planning requirements that anticipate customer demand are considered to be a push system, while kanban is considered to be a pull and just-in-time system (Pettersen & Sergerstedt, 2009). Traditional manufacturing strategies are driven by push systems, where the goal is to keep a large inventory of products according to customer forecasts (Naufal *et al.*, 2012). However, customer demand can change; therefore, such a system often creates an imbalance of stock between processes, leading to high levels of work-in-progress inventories, excess equipment, and surplus workers (Monden, 1983). Conversely, a pull production system is characterised by authorising the release of work based on system status (Hoop and Spearman, 1996), which prevents the need for inventories, leading to improved job performance (Koskela, 1992). Because the kanban technique is a pull system, it is possible to eliminate wastes, particularly those related to overproduction, work in progress (Productivity Press Development Team, 2002) and unnecessary stock, or to minimise waste (Low & Mok, 1999). Kanban can also be viewed as a visual device for managing and ensuring just-in-time production; it can be a simple and direct form of communication among all stakeholders (Ohno, 1988). A visual device is intentionally designed to share information at a glance, without requiring a written or an oral description (Galsworth, 1997). Finally, the kanban encourages continuous systematic improvements, making it a valuable tool to control and improve processes that are in the domain of all employees, and that are related to process control and the development of administrative, technical and operational project standards (Koskela, 1992).

8.3 The kanban technique

The kanban technique is an information system that harmoniously controls the production quantities of a process (Monden, 1983). According to Ohno (1988), the idea of the kanban emerged in the mid-1950s from American supermarkets, where goods bought in boxes were transported using a card containing information on the amount and types of goods purchased. When the purchase department received these cards, they could quickly replace the purchased goods on the shelves (Ohno, 1988). The kanban mainly uses a piece of paper within a rectangular vinyl envelope. This piece of paper contains information that can be divided into three categories: (1) pick-up information, (2) transfer information and (3) production information (Ohno, 1988). At Toyota, the kanban carries the information both vertically and laterally, within the Toyota factory, and between Toyota and cooperating firms (Ohno, 1988). The objective of using a kanban is to describe the quantity, time and location of particular parts during distribution, by providing the necessary information about the product, thus avoiding overproduction (Monden, 1983). The two types of kanban that are currently used are: (1) the transport kanban and (2) the production kanban (Productivity Press Development Team, 2002). Transport kanbans are used to either signal the need to replenish materials from a preferred supplier (supplier kanbans) or signal movement of parts or subassemblies produced within the factory to the production line (in-factory kanbans). Similarly, production kanbans describe whether to either initiate production (production-ordering kanbans) or communicate the need for machinery changeovers (signal kanban).

The main feature adopted by the kanban is that the store restocks only what has been sold, instead of using a system of estimated replenishment; this reduces defective inventories (Shingo, 1989). As stated by Ohno (1988), using the kanban technique, production workers start work by themselves, and make their own decisions with regard to overtime. Kanban also clearly indicates what must be done by managers and supervisors, supporting waste reduction and allowing for creative study and improvement proposals (Ohno, 1988). However, if used improperly, kanban is a technique that can cause various problems. Therefore, to employ kanban properly and skilfully, Ohno (1988) established functions and rules for kanban use, as follows:

1   Later processes pick-up the number of items indicated by the kanban in the earlier processes.
2   Earlier processes produce items in the quantity and sequence indicated by the kanban.
3   No items are made or transported without a kanban.
4   Always attach a kanban to all goods.
5   Defective products are not sent to their next process. The result is 100 per cent defect-free goods.
6   Reducing the number of kanbans increases their sensitivity.

In these six rules, Ohno (1988) highlights the importance of management commitment, adherence to the rules, the need to reduce worker resistance and convince workers that producing as much as possible is no longer a priority, and the need to shorten setup times and reduce lot sizes, so as to avoid excess inventory.

8.4 Research method

To describe the applications of kanban that can reduce waste in construction, empirical studies were conducted in two stages. In the first stage, an exploratory study was performed to identify current practices and opportunities for improvement in kanban implementation, based on assessment criteria in three construction projects. The second stage involved a pilot study that investigated the implementation of a kanban technique, given certain project conditions and using information gathered from the previous stage.

Four projects involved in the development and construction of residential and business buildings for the upper middle class were used in this study; all were located in the city of Salvador in the state of Bahia in Brazil. Table 8.1 shows the main characteristics of the projects studied.

8.4.1 Exploratory study

The evaluation assessments were performed in projects A, B and C. Data were collected through structured questionnaires that covered concepts and tools applied in the kanban technique, which have already been discussed, such as (1) mechanisms to reduce losses, (2) mechanisms to achieve continuous improvement, (3) tools to improve communication and transparency, (4) the concept of just-in-time and (5) pull production. The questions were evaluated as 'yes', 'no' or 'not applicable', from which a percentage of implementation for each major item was calculated. The structured questionnaire was applied through interviews conducted with the project manager, and through field visits, and photographs were taken to record the accuracy of the data collected.

Table 8.1 Main characteristics of the construction sites involved in the case studies

Project	Main characteristics	People involved in the study
A	A vertical construction site, containing one building with 16 floors. Transport kanbans were used for ceramic laying.	Production manager
B	A vertical construction site with two 15-floor buildings. Production and transport kanbans were used for mortar supply.	Production manager
C	A vertical construction site with two 15-floor buildings. Transport kanbans were used for ceramic laying.	Production manager
D	A vertical construction site with two commercial buildings: one with 15 floors, and the other with 23 floors. Production and transport kanbans were used for mortar supply.	Project manager, superintendent, foreman, two civil engineer trainees, plastering crew leader, person in charge of material control, and person in charge of mortar production

8.4.2 Pilot implementation study

A pilot study was developed in project D, with the support of an undergraduate student under the supervision of the principal author of this chapter.[1] Production and transport kanbans were used in this study for the mortar processes related to wall coating and floor regularisation tasks located at the hallways in one of the buildings. The production and transport kanban cards were made of plasticised paper, and contained information such as the quantity of mortar to be produced, the type of mortar to use, the delivery floor number and the delivery time. A kanban board was also created for this project. Two training meetings for the kanban technique were attended by the project manager, the superintendent, the foreman, two civil engineer trainees, the coating crew leader, the person in charge of controlling material supply, and the person in charge of producing mortar. Four control worksheets were developed for kanban implementation and monitoring, namely (1) a work package sheet, to be used by the foreman and the superintendent, (2) a material control sheet, to be used by the person in charge of controlling material supply, (3) a daily production control sheet, to be used by the crew leader in charge of the work package, and (4) a control sheet indicating the amount of mortar mixture to be produced, to be used by the person in charge of producing mortar. Examples of these control sheets are shown in Figure 8.1. Based on the collected data, the main indicators used in this study were the consumption of cement, the amount of material stock and the duration of the work package. Data were collected using worksheets, participant and direct observations, six formal meetings between the researchers and the production team, to discuss positive aspects

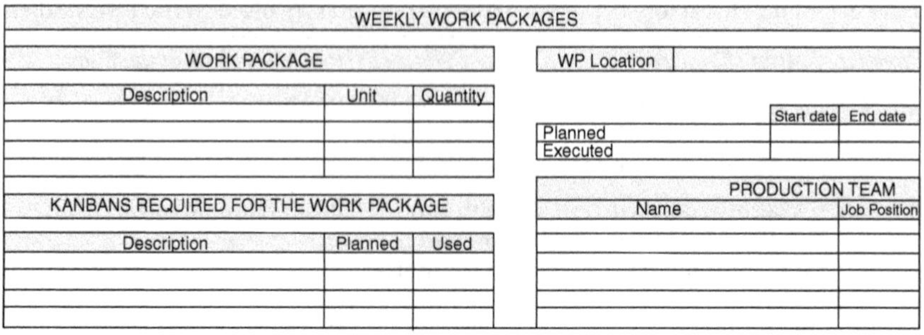

WEEKLY WORK PACKAGES						
WORK PACKAGE			WP Location			
Description	Unit	Quantity			Start date	End date
			Planned			
			Executed			
				PRODUCTION TEAM		
KANBANS REQUIRED FOR THE WORK PACKAGE			Name			Job Position
Description	Planned	Used				

Figure 8.1 Weekly work packages

of improvements of kanban implementation, and informal interviews with the field production team and structured questionnaires. Data analysis was also based on the criteria for assessing kanban implementation, which were developed in the first stage.

8.5 Good practices and improvement opportunities of the kanban technique in construction sites

Table 8.2 describes the kanban technique used in the three projects analysed, and Figure 8.2 shows a comparison of their use.

In terms of reducing material and time losses, good practices in each project were identified, including the use of kanban cards to control material stocks. At the end of each work day, each construction project submitted data to their stock system regarding the amount of material used during that day, to help establish and control the minimum stock levels necessary for the service. Despite this improvement, it was observed that none of the projects performed an in-depth study of the physical process flow, with the goal of reducing non-value-adding activities, such as transportation. Most of the construction site logistic plans were developed based on the experience gained by site engineers and foremen. Another problem observed was that no project used a tool to identify problems, or whether a worker was using more material than necessary. Therefore, none of the construction sites showed any concern for documenting problems, or taking action to improve their production processes. To achieve continuous improvement, projects A and C used an interesting systematic evaluation of their production teams that performed different services. In these two projects, the employees were trained to follow procedures, and different groups of employees had different targets to achieve. Inspections were then performed systematically, and each group was rewarded when their service was of a high quality and finished on time. Despite the faults detected by the quality systems, none of the construction sites sought to understand the causes of these failures. According to Ohno (1988), problems should be treated from the root of the cause, because only then

Table 8.2 A description of the kanban systems used in the different case studies

Project	Kanban system used	A brief description of the kanban system
A		In project A, a transportation kanban was used in the ceramic laying service. These kanbans worked as follows: cards were made by noting the materials required to perform the service in each room, as shown in the figure to the left; the quantity required of each material, its specification, the place of application and the total weight of the kit are also specified on each card.
B	Contrapiso Para Cerâmica Apto 903 7:00 Hs	Project B used a kanban system for only the mortar supply, while the ceramic blocks for building the brick walls were delivered without any control. For the mortar, the bricklayers received small cards identifying the mortar specification, where the mortar should be discharged, and the time it should take to be produced and transported to the job.
C		Project C used a kanban system for the transportation of the ceramic laying service. The idea came about as a result of the high number of personalised apartments. Card color coding provision scheme was made for floor cover, wall cladding, and adhesive mortar. The cards show information such as the amount of material required, the apartment to be plastered, the number of rooms planned and materials specifications.

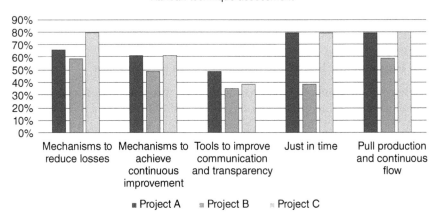

Kanban technique assessment

Figure 8.2 A comparison of the use of kanban technique in the three projects studied

can they be prevented from recurring. Communication and transparency were the main problems identified in the projects studied. None of the construction sites provided design and planning information to the workers, and it was noted that the workers were not encouraged to learn about the project, and they had difficulties in performing the processes according to the plans and the design. If there was ever any doubt, the person in charge had to be located and brought to the site, which wasted time while the workers waited for the information to arrive. It also created dependence on the construction site managers. A good practice identified in project B was the use of flags as a way of communicating the presence of a material (use of a green flag), or the need for more material (use of a red flag) (see Figure 8.3). Another good practice identified in project C was the use of a board showing performance measurements related to service quality, which was placed in the lunchroom.

When analysing the concept of just-in-time, none of the construction sites had a schedule plan for material delivery from suppliers. According to the interviewees, the difficulty lies with suppliers that cannot perform on-time deliveries, which ends up forcing the construction site to accumulate stock. Considering pull production, all projects generally sought to keep their production balanced by establishing a production rate for each task, and aligning the work packages of the kanban with weekly work planning. However, none of the projects performed monitoring of the weekly plan, or tried to record the causes of failure of some activities. Thus, there is no feedback regarding accomplishment of the work packages detailed on the kanban cards. The results of this exploratory study show that the kanban implemented in the projects analysed were being used only in a very simple way. It was observed that most of the principles and concepts were not well understood by the construction managers. Interviews with engineers, trainees, and persons in charge of controlling the material supply showed that the system helps and facilitates material control. However, it seems that there is still a lack of theoretical knowledge regarding the concepts and principles behind the tools among the production teams involved.

8.6 Implementation of the kanban system on a job site

An implementation of the kanban technique was investigated in project D, which used the concepts and tools for system monitoring in a more detailed way. This project had already implemented some lean construction practices, such as the LPS (Ballard, 2000), which produced more stable production, which is an important condition for the introduction of kanban ideas. The starting point of the implementation was the definition of production and transport kanbans. Eleven types of mortar mixture were identified for wall coating and floor regularisation tasks related to hallways; a colour card was assigned to each mixture type. A kanban panel was built using port timecards that already existed on the construction site. The

Figure 8.3 The use of flags for communication about materials

columns were subcategorised according to material delivery time, and were set up by the production team, by considering the regular time that the material was usually requested. A list of all the types of mortar mixture was made available to the person in charge of producing mortar, and placed near the mortar mixer. Figure 8.4 depicts the kanban cards and boards used in project D.

The first week of implementation included only basic features of the kanban technique, such as the process of material requests using the cards and the kanban boards, to make the production team more familiar with the system. Thus, initially the kanban was mainly used as a communication tool. The following lessons were learned from this basic implementation stage:

- Productivity improvement. Before implementation, the activity of floor regularisation of the bathroom in the hallways took two days. One of the reasons for this was that the material was not available for the coating team at the beginning of the shift. Therefore, a long wait time was identified between ordering of mortar by the labourers and production and delivery of mortar by the person in charge of producing mortar. After one week of implementation, one day of cycle time reduction was observed, mainly because the material was available in the quantity required, and when the team required it.
- Production imbalance and better transparency. During the first week, the kanban panel showed some empty time slots. It was observed that the crew leader requested mortar only once per day; however, this caused mortar quality problems due to the limited time of its use. This incorrect practice was shown clearly on the kanban panel, and was solved after a new orientation meeting was held between the superintendent, the foreman and the crew leaders.
- Incorrect use of mortar mixture. It was noted that the number of packages of cement used in the first week was significantly more than planned. Searching for the root cause of this problem, it was identified that despite the availability of a board with a list of mortar mixtures at the mixer space, the person in charge was producing mortar according to his own knowledge from a previous construction project. Therefore, new instructions were given to this worker to resolve this important technical issue.

In the second stage of the implementation, the goal was to use kanban for production control, in addition to communication; thus, monitoring of worksheets was introduced. Data concerning the number of packages of cement planned and actually used, as well as the percentages of waste, were collected. The data are presented in Table 8.3.

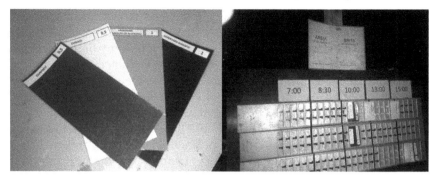

Figure 8.4 Kanban cards and boards used in project D

Table 8.3 Consumption of cement during kanban implementation

Activity	Number of packages of cement		Cement wasted (%)
	Planned	Used	
Week 1 – floor regularisation	42	78	86
Week 1 – wall coating	17	23	35
Week 2 – wall coating	4.5	9	100
Week 3 – wall coating	25	30	20

The findings show significant deviations in cement consumption during implementation. The production team identified a change in the design of the floor regularisation that was not communicated to the person in charge of project cost estimating and monitoring. With regard to the wall coating, the data showed that the root cause of the increase in cement use was use of a different mortar mixture than that which was previously specified, and a higher thickness of mortar than that previously specified. In terms of completion of planned work packages, all packages were completed according to plan, even with some external interference during the week; this showed that the kanban had a positive influence on the completion of work packages on time, thereby increasing productivity. The main lesson learned in this stage was that the entire production team, including the foreman, the crew leader, the person in charge of controlling material supply, the person in charge of producing mortar, and the managerial team, should be involved in implementation, monitoring, identifying the root cause of problems, and taking actions to resolve the problems. One piece of evidence of this engagement is the informal report of the person in charge of producing mortar, who affirmed that using the kanban technique eliminated the number of last-minute orders of mortar. For the project manager, the use of the kanban reduced material waste and identified the problems of using a different mortar mixture in the field, delays in material deliveries and delays in the execution of work packages, thereby allowing immediate actions to be taken. Considering the principles behind the kanban technique, this pilot implementation identified important improvements over the previous projects studied. With regard to waste elimination, the use of kanban and the worksheets for the monitoring thereof supported the identification of cement waste and its root causes, thereby reducing overproduction and material waste; this monitoring also contributed to continuous improvement. The kanban boards and cards, the list of mortar mixtures and the indicators collected provided better communication and greater transparency of information for the production and managerial teams, which helped to identify waste in processing and waiting time. In addition, the kanban technique achieved effective pull production and just-in-time production, because mortar was produced based only on the demand of the workers, which avoided the use of inventory and prevented transport wastes and additional waiting time.

8.7 Necessary conditions and guidelines for implementing kanban

Based on these empirical studies, two necessary conditions – namely the planning and production control for production stabilisation and the planning of the construction site logistics for appropriate physical flow – were identified.

Planning and production control. Preparing the production planning and control can be based on the LPS (Ballard, 2000) and is the first step to achieve an effective production process where goals can be established. The design and implementation of the kanban technique is strongly related to the hierarchical levels of production control. In the master schedule, macro tasks are defined, which is important for identifying potential tasks for kanban implementation, and for supply relations between material purchasing, equipment and labour. In look-ahead planning, the primary objective is to analyse constraints, to determine what must be done to make a task ready to be executed. This is a critical level of design for a kanban technique. Material supply, labour force, equipment, final design drawings and physical flow are important constraints to be considered. Based on the work packages already coordinated and established in the look-ahead plan, kanban cards may be designed according to the number of tasks defined, as well as the mechanism for system control. It is essential to consider the core concepts, such as the concept of just-in-time, transparency, the pull system architecture, continuous flow and continuous improvement in the kanban design stage. In weekly work planning, work packages are defined, and their execution is often monitored using the Plan Percent Complete (PPC) measure, which is defined as the ratio of completed work packages to the total scheduled activities (expressed as 100 per cent). It is important to remember that during construction, monitoring of the PPC should be performed so that the causes of non-fulfilment of planned activities can be identified. At this level, the kanban, as well as mechanisms to reduce and control waste, should be implemented and associated with the LPS tools.

Planning of job site logistics. Logistics is part of material management because it determines delivery of goods to the site, location of a space for offloading, location of a space for storage, protection of materials and movement of materials to construction areas (Low & Mok, 1999). A well-planned construction site layout is essential in order to reduce waste and non-value-adding activities in a construction project. A study of the physical flow of the project, which considers material and labour flows (Alves, 2000), including distances, waiting times and transport activities, is important for the implementation of a kanban. In construction site planning, it is also interesting to analyse all stages of the construction of the building, from its foundations to its completion, using long-range planning systems, to define the amount of material stocks, and their location on the construction site. Long-term planning with budget considerations is directly linked to choice of equipment, and is therefore essential during the design process. Yield data and time may affect the choice of equipment, or quantities allocated. A set of practical guidelines was established to support the design, implementation and monitoring of a kanban system in construction projects, so as to reduce waste.

Identification of the potential processes for kanban use. As mentioned, identification of the potential processes to apply the kanban system should be performed, based on the work packages of the master schedule and the look-ahead plan. It is important to identify tasks that require material management, and the types of kanban required. Examples of materials that demand transport kanbans include bricks, blocks, ceramic tiles, drywall frames, doors, kitchen and toilet accessories, and plaster, cement and mortar packages. Mortar, concrete and prefabricated components are examples of materials that can demand production and transport kanbans, because they are produced on-site and then transported based on client demand.

Information availability about the processes selected. For the design and planning of a kanban, detailed information about the required tasks is necessary. Essential information includes final drawings, construction activities quantity take-off and materials to be used, labour productivity rates, the sizes of available crews, availability and requirements of equipment, site layouts and floor layouts.

Kanban devices and tools. Based on the detailed information of the process in which the kanban will be applied, cards, panels and other visual devices may be used. There is no specific model for this, and the production team can be creative when preparing devices using the available resources so as to reduce costs. However, it is important to guarantee that the devices, particularly the cards, provide essential information to implement the pull system, such as specifications of the part to be produced or transported, the quantities to be produced or transported, the process responsible for the production and use of a specific part (i.e. the previous and subsequent processes) and the storage location (Monden, 1983). In addition, worksheets for production and material management may be created, such as the ones used in the pilot study presented in this chapter; these can include work package sheets, material control sheets, daily production control sheets and stipulations of the amount of mortar mixtures required. This will provide data for measuring inventory wastes, material consumption wastes, delays in material delivery and task completions.

Staff training. After planning the system, the implementation should consider training the entire team – including production managers, the persons in charge of managing material supply and producing material and the workforce – on the concepts, principles and rules of the kanban technique. To make everyone feel part of the system, it is important that everyone participates in the creation of the cards, as well as the control mechanisms, such as the kanban boards, control worksheets and control charts.

Implementation in steps. A good practice that was identified in the pilot study was the implementation of the kanban in steps. In the first weeks, the goal is to make the production team more familiar with the technique, so as to improve communication between the different people involved. In the second stage, the goal of production control may be added to the technique using a control mechanism to support identification of wastes, and other technical issues, thereby enabling workers to start work by themselves and make their own decisions with regard to overtime (Ohno, 1988), and enabling the managerial team to take action to reduce waste.

8.8 Summary

This chapter examined some applications of the kanban technique in construction projects reported in the literature; empirical studies were used to show that the kanban is a technique that may help to minimise problems concerning several types of wastes, such as overproduction, inventory, transportation, excess of material consumption and waiting time at the construction site. In addition, it contributes to improved communication and transparency, better labour productivity, and, as a consequence, keeps a project on time and on budget. The first contribution of this chapter is the establishment of evaluation assessment criteria for kanban implementation, which include mechanisms for reducing waste, mechanisms for achieving continuous improvement, tools for improving communication and transparency, the concept of just-in-time and production and pull streaming.

When considering implementing kanban, companies must first formulate a plan of how process steps will occur, and then consider training employees on how the technique functions, its importance and how to resolve problems. Therefore, there are two necessary conditions for kanban implementation: (1) the adoption of production planning control for production stabilisation and (2) job-site logistics planning for appropriate physical flow. In addition, practical guidelines can support the design and implementation of a kanban technique, such as identification of the potential processes for kanban, information availability regarding the processes selected, kanban devices and tools, staff training and implementation in steps.

Notes

In developing this chapter, the authors have drawn on papers published at an IGLC conference (Burgos and Costa, 2012). The author gratefully recognises the International Group for Lean Construction in this regard.

Endnote

1 Renan Ribeiro Lima Menezes is the author of the undergraduate dissertation in BSc (Civil Engineering), titled *Guidelines for effective implementation of Kanban in construction sites* (in Portuguese) (Menezes, 2013). The data related to the pilot study used in this chapter are part of his dissertation.

References

Alves, T.C.L. (2000), Diretrizes para gestão dos fluxos físicos em canteiros de obra: proposta baseada em estudo de caso. Master Dissertation. Núcleo Orientado paraa Inovação da Edificação Programa de Pós- Graduação em Engenharia Civil, Universidade Federal do Rio Grande do Sul, Porto Alegre (in Portuguese).

Arbulu, R. (2009), Application of integrated material management strategies. In: W.J. O'Brien, C.T. Formoso, R. Vrijhoef & K.A. London (eds), *Construction Supply Chain Management Handbook*, CRC Press, Boca Raton, FL.

Arbulu, R.J., Ballard, G. & Harper, N. (2003), Kanban in construction. In: *11th Annual Conference of the International Group for Lean Construction (IGLC)*, Blacksburg, VA, 22–24 July.

Ballard, G. (2000), The Last Planner System of production control. PhD thesis, University of Birmingham, UK.

Barbosa, G., Andrade, F., Biotto, C. & Mota, B. (2013), Implementing lean construction effectively in a year in a construction project. In: *21st Annual Conference of the International Group for Lean Construction (IGLC)*, Fortaleza, Brazil, 29 July–2 August.

Burgos, A.P. & Costa, D.B. (2012), Assessment of Kanban use on construction sites. In: *20th Annual Conference of the International Group for Lean Construction (IGLC)*, San Diego, CA, 18–20 July.

Forbes, L.H. & Ahmed, S.M. (2011), *Modern Construction Lean Project Delivery and Integrated Practices*, CRC Press, Boca Raton, FL.

Galsworth, G.D. (1997), *Visual Systems: Harnessing the Power of a Visual Workplace*, AMACOM, New York.

Heineck, L.F., Rocha, F.E., Pereira, P.E. & Leite, M.O. (2009), *Building Lean Collection: Building with Lean Management*, Publisher Graphic Expression (in Portuguese), Fortaleza, Brazil.

Hoop, W.J. & Spearman, M.L. (1996), *Factory Physics: Foundations of Manufacturing Management*, Irwin, Chicago, IL.

Jang, J.W. & Kim, Y.-W. (2007), Using Kanban for construction production and safety control. In: *15th Annual Conference of the International Group for Lean Construction (IGLC)*, East Lansing, MI, 18–20 July.

Junior, M.L. & Filho, M.G. (2010), Variations of the Kanban system: Literature review and classification. *International Journal of Production Economics*, Vol. 124, pp. 13–21.

Kemmer, S.L., Saraiva, M.A, Heineck, L.F.M., Pacheco, A.V.L., Novaes, M. de V., Mourão, Carlos A.M.A. & Moreira, L.C.R. (2006), The use of andon in high rise building. In: *14th Annual Conference of the International Group for Lean Construction (IGLC)*, Santiago, Chile, 25–27 July.

Khalfan, M.M.A., McDermott, P., Dickinson, M.T., Li, X. & Neilson, D. (2008), Application of kanban in the UK construction industry by public sector clients. In: *16th Annual Conference of the International Group for Lean Construction (IGLC)*, Manchester, UK, 16–18 July.

Koskela, L. (1992), Application of the new production philosophy to construction. Technical Report 72, Stanford University Center for Integrated Facility Engineering (CIFE), Stanford University.

Low, S.P. & Chan, Y.M. (1997), *Managing Productivity in Construction: JIT Operations and Measurement*, Ashgate, Aldershot.

Low, S.P. & Choong, J.C. (2001), A study of the readiness of precasters for just-in-time construction. *Work Study*, Vol. 50, pp. 131–140.

Lu, D. J. (1985), *Kanban Just-in-Time at Toyota: Management Begins at the Workplace*, Productivity Press, Cambridge, MA.

Menezes, R.R.L. (2013), Guidelines for effective implementation of kanban in construction sites. Undergraduate dissertation in Civil Engineering BSc, Federal University of Bahia, Bahia (in Portuguese).

Monden, Y. (1983), Production without inventory: A practical approach to the Toyota production system. Institute of Industrial Engineers, Norcross, Georgia.

Naufal, A., Jaffar, A., Yusoff, N., Hayati, N. (2012), Development of kanban system at local manufacturing company in Malaysia: Case study. *Procedia Engineering* Vol. 411, pp. 721–726.

Ohno, T. (1988), *Toyota Production System: Beyond Large-scale Production*, Taylor and Francis Group and Productivity Press, Portland, OR.

Productivity Press Development Team (2002), *Kanban for the Shopfloor*. Productivity Press, Portland, OR.

Shingo, S (1989), A Study of the Toyota Production System: From an Industrial Engineering Viewpoint, Productivity Press, Portland, OR.

9 Use of andon in a horizontal residential construction project

Clarissa Biotto, Bruno Mota, Lívia Araújo,
George Barbosa and Fabíola Andrade

The andon is a visual management tool used mainly in manufacturing systems to highlight the status of operations. In understanding the difficulties imposed by this kind of production, it is necessary to adapt the concept and physical structure of this tool to construction sites. An andon tool was implemented in a 55-hectare area of a large horizontal residential project of more than 800 apartments in Brazil, in order to indicate problems in each workstation for the project management team through audible and visual signals using a television screen. The tool was adapted to a touch-screen terminal near the workstations and it was used by 900 workers, each one having an individual identification number. The workers have to log-in at the andon terminal and indicate the actual status of the operation they are executing. As a result, the andon increased operational transparency relative to problem identification at the front-line of operations and it helped to resolve interruptions pertaining to: lack of material, problems with manpower, design documentation and safety. Also, the andon improved workers' sense of responsibility since they were requested to anticipate problems that could cause work stoppages. The andon tool presented in this chapter can guide other construction companies on how to implement this tool at large construction sites.

9.1 Background

Developed for the Toyota Production System (TPS), the andon has been widely used in manufacturing plants to improve product quality (Li & Blumenfeld, 2006). However, its application in the construction industry is very limited owing to its special kind of production, which is very different from that found in manufacturing. According to Bertelsen and Koskela (2004), construction is a turbulent kind of production that 'makes one-of-a-kind products and does so at the site by cooperation within multi-skilled ad-hoc teams'. These characteristics hamper the adoption of andon on construction sites owing to the difficulties in installing the andon devices in every workstation or site office. Even so, as detailed by Kemmer (2006), there are cases of successful application of this tool in some high-rise buildings where the andon system uses the electrical infrastructure of the building to enable workers to request help and avoid an interruption to their activities. However, this strategy may be expensive in horizontal residential projects, where large areas have to be covered and wired. This chapter focuses on a situation whereby a project management team faced the challenge of using andon at its own construction site, which has 99 apartment blocks spread over 55 hectares. In this project, the administrative team and the employees were already using the concepts and the tools of lean construction, which facilitated the implementation of the andon.

9.2 Andon

The andon is the basis for one of the TPS main pillars – jidoka, which means 'machines with human intelligence'. It is a management tool of visual control that shows the operation status in a workstation (Lean Enterprise Institute (LEI), 2008). The idea is to give workers the autonomy to send a signal seeking help from their supervisors and stopping the production flow when there is a problem in their workstations (Shingo, 1989; Liker, 2004; Liker & Meier, 2006). That problem is immediately communicated to the team, and the group leaders are the ones responsible for checking exactly where the problem is (Liker & Meier, 2006). The most important action is the solution to this problem (Shingo, 1989). It is the basis of a culture of stopping the work to solve problems in order to achieve quality (Liker, 2004; Liker & Meier, 2006). In many companies, when trying to implement an andon, workers usually have difficulty in admitting they need help. And if resentment develops among workers or leaders, the andon will be ineffective (Liker & Meier, 2006). The andon tool includes an audible alarm and a visual light to indicate the location of the problem (Liker & Meier, 2006). It describes the actual status of the production – for example, how many machines are operating, an abnormality such as a quality problem and defective tools, among others (LEI, 2008). When implementing an andon, the problems will no longer be hidden, but will be detected and can be fixed (Li & Blumenfeld, 2006).

9.3 Application of andon in the construction industry

Previous applications were made in another construction company in Fortaleza, Brazil, that had lean implementations for almost ten years. In a high building, the andon is used together with kanban cards and the heijunka box. Each workstation in the building has an andon panel composed of three triggers switches: green, yellow and red (Kemmer *et al.*, 2006, Valente, 2011). These switches turn on the light emitting diode (LED) on the control panel installed at the management office so the engineering team can see the production status. The green button turns on a green LED to indicate that the activities are taking place normally. The yellow one turns on a yellow LED to indicate that an activity will be interrupted in the next 30 minutes. The red one indicates that production has stopped completely.

Every morning, when workers arrive at the workplace, they press the green button on the andon panel. If they know that some materials, tools or information are missing, or that any detail has been left off the drawings, or that any interruption could stop their job, one of the workers will press the yellow button and wait for the supervisor to get in touch by radio to solve the problem. But if the team stops the work, and the red button is pressed, the supervisor gets in touch with them immediately and tries to fix the problem (Kemmer *et al.*, 2006; Valente, 2011). The reasons for the interruption of work packages could be, among others, labour training, production system capacity, planning or supplier delay. All these reasons should be determined after applying the Five Whys technique (Kemmer *et al.*, 2006). In projects with this kind of andon, the work interruptions are reduced significantly. Also, the transparency is enhanced, improving the communication among the management team and the employees. The main difficulties in keeping this manual system running pertain to: getting the workers to press the buttons on the andon panel depending on the status of their activities; getting management teams to record the reasons for production stoppage; and getting management to record the crew that has stopped the work. Also, it is easier to use the andon in an LED control panel when the structure of the building has already been built, because after that the workstations are physically defined.

9.4 Case study description

This study reports on the implementation of the andon at a construction company founded in 1980 at Fortaleza, Brazil. It has delivered 100 buildings, including commercial and residential projects, flats and others, all in prime areas of the cities. The company has had a total quality programme since 1998, ISO 9001/2000 certification since 2004 and has practised lean implementation in its sites and offices since 2013. The case study project in this chapter is a condominium resort in an area of 553,545.74 square metres. The project began in 2010 and has a term of ten years. During this period, an extensive leisure area with swimming pools, barbecue area, golf courses, sports and other facilities will be constructed, with 82 houses and 99 apartment blocks. Figure 9.1 presents a perspective of the project.

9.4.1 Activities incidental to andon implementation

The main difficulty in implementing the andon tool on this site was the extensive size and the considerable number of apartment blocks in the project. The difficulty was installing an andon panel in each one. If it were

Figure 9.1 Plan of the project and a proposed apartment block

to be made, in total it would be necessary to install 744 trigger switches and more than 100 kilometres of electrical cable on the site to connect all the workstations to the control panel in the main office. Therefore, the engineering team decided to use information technology to transfer the data around the site using an existing Wi-Fi infrastructure, derived from one year of lean construction implementation on the project (Barbosa *et al.*, 2013). Owing to this, the andon is a continuation of the implementation of different lean concepts, tools and techniques. Therefore, all the programming language and infrastructure remained the same. The engineering team thus added the andon at the lean terminal where the workers could trigger the device when it is required. In brief, the andon system development activities include:

- Development of the tool. This step included the scope definitions and its adaptation to the construction site nature. It was programmed in the Delphi language, using SQL Server 2008 database and a web service to send SMS messages to employees' cell phones. This step required the purchase of a lean terminal touch-screen. The andon was connected to the main lean system software through the site wireless network, which supports the production planning and control (Barbosa *et al.*, 2013).
- Testing and improvements. During a 15-day period, the andon was tested in only one terminal: one apartment block on the site. During this step, it was necessary to train one employee on how to use the andon to be next to the lean terminal to explain the use of it to the others. This phase occurred during the period from the end of September to the middle of October 2013.
- Consolidation. After the trial test, the andon system was up and running. The tool was evaluated and the results were measured according to some indicators, with data from October 2013 to February 2014.

9.4.2 Development of the andon

The andon tool was installed in three lean terminals spread throughout the main streets of the site. They are on the ground floor of the apartment blocks and their locations are indicated by signposts on site (Figure 9.2). The andon has to be activated through the lean terminal which contains other tools used by employees (Figure 9.3).

The activation can be done by a worker that intends to signal that his crew is stopping production. Then, he needs to log into the tool, choose the andon screen, inform his workstation, give the reason why they are going to stop and confirm his activation. There are four categories of activity stoppages: material, crew, design and safety. The engineering team receives an alert on the site office screen and they have ten minutes to solve problems relating to safety and 30 minutes for others categories before production stops. The administrative assistant identifies the activities that could be

Figure 9.2 Lean terminals location and signalisation on construction site

Figure 9.3 Lean terminal where workers can access the andon device

interrupted as well as the reasons. He is responsible for solving material, manpower, design and safety problems on the site. Then, if he is not able to solve them by himself, he asks safety technicians (for safety problems) or engineers for help. Most of the andon's problems are simple to solve, and the engineers are not notified.

9.4.3 Outcome of the use of andon in the case study

Some andon indicators can be produced easily owing to the automatic collection of data. For this particular project, data were collected over a period of five months – from October 2013 through to February 2014. The total andon activation in this period was 187. This number is broken down into crew, material, design and safety problems. It was observed that material is the main reason for andon activation by the workers, representing 82 per cent. The distribution of andon activation during these months shows the intensive use of this tool in the first two months of its implementation: 62 activations in October and November, 2013. In the following months, this number progressively decreased because of two possible reasons. One, the engineers may have learned how to programme the material requisitions for their scheduled work. Two, the workers may have decided to make extensive use of the kanban cards for material requisition. Another indicator was the andon activation to determine schedule deviation. During five months, 92 per cent of all andon activation was to request some resource for the same workstation and activity that had been scheduled in the commitment plan. Schedule deviations accounted for 8 per cent of all andon activation in this period. The efficiency of the engineering team in solving problems before the activity interruption was evaluated (Figure 9.4). It can be seen that the safety category was the most difficult to solve in a few minutes. Only 59 per cent of all requests in the andon system were resolved before the activity stopped. Of all andon activations, it was possible to analyse the time deactivation average in the four categories. The material category took longer to turn off the andon: 37 minutes on average for solving the problem; 48 minutes on average for activities that have stopped.

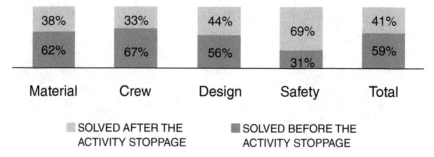

| Material | Crew | Design | Safety | Total |

SOLVED AFTER THE ACTIVITY STOPPAGE · SOLVED BEFORE THE ACTIVITY STOPPAGE

Figure 9.4 Percentage of activities solved before and after stoppage

Other indicators deal with labour. An indicator shows that only 3 per cent of all andon activations were performed by outsourced labour, and 97 per cent from within the main workforce. The reasons could be because of the good performance of the outsourced company, or if there was a lack of some resource, only the outsourced company could solve the problem with its labour, and the andon activation was not necessary. Observations on the site also illustrate the fact that the professionals who used the andon most were the masons, at 48 per cent.

9.5 The relation of the use of andon to waste control

By using the andon, it is possible to identify and control wastes on the construction site. After andon activation, as the administrative team tries to resolve the problem indicated by the workers, they are also avoiding common types of waste: motion, processing, inventory, conveyance and waiting, all of which are described in Ohno's list. In addition, wastes by making-do, which is described by Koskela (2004) as the eighth waste of construction, task diminishment (Koskela & Bølviken, 2013) and breaks in tasks sequence. For instance:

- The waste by motion occurs when workers carry out unnecessary movements such as looking for parts, tools, documents, among others, during the execution of an operation. Also, this waste can happen during the movement of workers between different workstations owing to a lack of planned activity sequence.
- The waste by processing is pointed out by Shingo (2005) as a waste related to the processing speed, processing method and the need for processing something. It may be caused by a lack of design information, a lack of process standard, a lack of adequate training of labour, among other factors. It is a waste characterised by unnecessary work or an inefficient way of undertaking an activity that in turn increases the process cycle time.
- The waste by making do occurs when a task is executed although all preconditions have not been met. It can directly affect the quality of the service which can both generate rework in the next activity and wastes by processing, or it can increase the safety risks of workers due to the lack of some safety equipment required to perform the activity. Thus, the generation of value is jeopardised in terms of the internal clients (next labour team) and may be reflected in the quality of the product as perceived by the final client (user).
- In the case of task diminishment, the task did not comply with specification and quality standards. It can be caused by the making do, and this task can remain uncorrected until the following activity starts. This could result in the need to reallocate other resources in the future to complete the task according to specification or generate informal

activity packages to finish the task. This diminished task can therefore delay the start of the next activity or cause rework and processing wastes.

9.6 Improvements linked to andon use

As it is a computerised tool, the data collection is automatic. When the andon is triggered, an alert appears at the screen of the engineering office with a countdown showing by when the problem must be resolved. From this moment, all data are registered in the tool, which improves the communication speed and makes the activity status transparency for all engineering teams. In this andon the worker is the one who indicates why he is going to stop the activity. This fact decreases the time required to solve the problem, because the engineer does not have to search for the crew: he or she only has to ask them the reason for the stoppage. As the worker needs to register his job or location, the andon calculates automatically where the worker should be according to the commitment plan and where the worker actually is. Thus, it is possible to identify schedule deviations that justify a lack of resources to perform a scheduled activity. It is also possible to identify informal activities that generate waste by making do. The most important benefits that the andon tool enabled are fixing problems and improving commitment to planning quality by the engineering team. Also, a very expensive resource such as manpower will not be idle owing to supply problems, and the wastes of making do can be reduced.

9.7 Difficulties linked to andon use

The difficulties encountered in the andon implementation at the site are related to employee engagement. Training must always take place for them to be updated because they must understand the importance of notifying the engineering team when they are going to stop or when they have already stopped production. The opposite also happened when, early in the implementation, there were some workers who triggered the andon without having made their requisitions through the kanban. One limitation of this andon tool is the employees' reluctance to trigger the andon to signal a green production situation that indicates they are working in the right place with all the necessary resources. Currently, the andon only indicates a yellow or red situation of production, and that is the reason the engineering team does not know whether the workers are in the correct workplace as scheduled.

9.8 Summary

There are several benefits associated with the andon tool in managing the supply of building blocks on the construction site. The information transparency led to improvements to the plan and decreased the number of stoppages in production. The andon is a tool for revealing flaws in medium- and short-term planning. First, it is necessary to remove all restrictions related to the execution of the activity; and second, there should only be scheduled activities for which there are no pending constraints, avoiding the consequent effects of variability and uncertainty in the production system. Use of the andon enables alerts about planning flaws, which in turn facilitates improvements in planning. There is a considerable increase in transparency. However, the andon faces problems with the employees' lack of willingness to trigger the device and obtaining the real production status. Nevertheless, it was an adaptation of the andon concept to an extensive construction site that supported the kaizen for production planning and control.

Note

In developing this chapter, the authors have drawn on papers published at an IGLC conference (Biotto *et al.*, 2014). The author gratefully recognises the International Group for Lean Construction in this regard. The authors also thank the directors of the Colmeia Construction Company and SIPPRO consultants' for the opportunity to develop this chapter.

References

Barbosa, G., Andrade, F., Biotto, C. & Mota, B. (2013), Implementing lean construction effectively in a year in a construction project. *Proceedings for the 21st Annual Conference of the International Group for Lean Construction*, Fortaleza, Brazil, 29 July–2 August, pp. 1017–1026.

Bertelsen, S. & Koskela, L. (2004), Construction beyond lean: A new understanding of construction management. In: *Proceedings for the 12th Annual Conference of the International Group for Lean Construction*, Elsinore, Denmark, 3–5 August, pp. 1–11.

Biotto, C., Mota, B., Araujo, L., Barbosa, G. & Andrade, F. (2014), Adapted use of Andon in a horizontal residential construction project? In: *22nd Annual Conference of the International Group for Lean Construction*, Oslo, Norway, 23–27 June, pp. 1295–1306.

Kemmer, S.L., Saraiva, M.A., Heineck, L.F.M., Pacheco, A.V.L., Novaes, M.V., Mourão, C.A.M.A. & Moreira, L.C.R. (2006), The use of andon in high rise building. In: *Proceedings for the 14th Annual Conference of the International Group for Lean Construction*, Santiago, Chile, 25–27 July, pp. 575–582.

Koskela, L. (2004), Making-do: The eighth category of waste. In: S. Bertelsen & C.T. Formoso, *12th Annual Conference of the International Group for Lean Construction*, Helsingør, Elsinore, Denmark, 3–5 August 2004.

Koskela, L. & Bølviken, T. (2013), Which are the wastes of construction? In: C.T. Formoso & P. Tzortzopoulos, *21st Annual Conference of the International Group for Lean Construction*, Fortaleza, Brazil, 31 July–2 August, pp. 3–12.

Lean Enterprise Institute (2008), *Lean Lexicon: A Graphical Glossary for Lean Thinkers*, Lean Enterprise Institute, Cambridge.

Li, J. & Blumenfeld, D. E. (2006), Quantitative analysis of a transfer production line with andon. *IIE Transactions*, Vol. 38(4) pp. 837–846.

Liker, J.K. (2004), *The Toyota way: 14 Management Principles from the World's Greatest Manufacturer*, McGraw-Hill, New York.

Liker, J.K. & Meier, D. (2006), *The Toyota Way Fieldbook: A Practical Guide for Implementing Toyota's 4Ps*, McGraw-Hill, New York.

Shingo, S. (1989), *A Study of the Toyota Production System from an Industrial Engineering Viewpoint*, Productivity Press, Cambridge.

Shingo, S. (2005), *A Study of the Toyota Production System*, CRC Press, Boca Raton, Fl, London and New York.

Valente, C.P. (2011), Lean monitoring and evaluation on a construction site: A proposal for lean auditing. (In Portuguese: Acompanhamento e avaliação lean em um canteiro de obras: uma proposta de auditorias lean. Monografia [graduação] – Universidade Federal do Ceará, Departamento de Engenharia Estrutural e Construção Civil, Curso de Engenharia Civil, Fortaleza.)

10 Wastes and *genchi genbutsu*

The importance of 'go and see for yourself' to project value

Bolivar A. Senior and Brad Hyatt

An essential element of lean management, and particularly of lean construction, is the insight that the best way to get a meaningful understanding of a problem is personally going to the place where action is taking place to observe the situation. This principle is called *genchi genbutsu* in lean management, and the management tool of deliberately observing the situation at hand is a *gemba* walk. This chapter discusses the origin, purpose and implementation of *genchi genbutsu* and *gemba* walks. The chapter describes a case study in a higher-education setting. *Gemba* walks played a breakthrough role in developing countermeasures to problems faced in an undergraduate course laboratory when productivity levels in the lab sessions did not seem to be improved by the Last Planner System (LPS). The chapter concludes with a discussion of the relevance of going to *gemba* for problem solving in general.

10.1 Background

General Dwight Eisenhower said that 'Farming looks mighty easy when your plough is a pencil, and you're a thousand miles from the corn field' (Eisenhower Presidential Library, n.d.). In other words, be there if you want to truly understand your project. Lean construction (LC) recognises the wisdom of this adage. The Last Planner System (LPS) (Ballard, 2000), for example, is centred on the input of field personnel who are in the best position to understand the problems and opportunities of the project. LC, in turn, aligns with the principles of the Toyota System and lean management in general, which also acknowledges the importance of personal presence to assess and provide solutions to management problems. This chapter discusses the 'go and see for yourself' approach to management as viewed by lean management, beginning with a description of its origins, characteristics and implementation. This general introduction is followed by a case study set in a higher-education environment. The case study discusses how the 'go and see for yourself' approach was important for the elimination of waste and improving the productivity of a small shed construction project in the context of a construction management course that introduced LPS.

10.1.1 Concepts and vocabulary: *genchi genbutsu*, *gemba* and going to *gemba*

Jeffrey Liker recounts in his pivotal book, *The Toyota Way* (2004), that when he asked Toyota employees who he interviewed what distinguished the Toyota Way from other management approaches, 'the most common first response was genchi genbutsu', regardless of the employee's position in the company. Toyota Motors identifies *genchi genbutsu* as one of the four high-level principles in the internal document used for employee training. In Japanese, *genchi* means 'actual place' and *genbutsu* means 'actual thing'. The combination of the two words is frequently translated as 'going to the place to see the actual situation for understanding' (Liker, 2004). The term *genchi genbutsu* is, thus, richer than the two words in isolation.

Gemba is another key word for lean management. It can be translated from the Japanese as 'the real place' or 'the actual place'. It is a relatively common word in Japan. After an earthquake, for example, an on-site TV crew may say that they are 'reporting from the gemba', meaning to be reporting from where the action is (Imai, 1997). In lean management, *gemba* refers to the place where value is actually created, such as the shop floor in a car manufacturing assembly. The *gemba* of a construction project from its owner's perspective are the points at which the project is actually built, all other places and steps being of importance to the contractor. *Gemba* is increasingly used as synonymous to *genchi genbutsu*, especially when used as 'going to *gemba*'. This chapter uses 'going to *gemba*' and '*genchi genbutsu*' interchangeably, both referring to going to the source to find the facts to make correct decisions, build consensus and achieve goals (Toyota Material Handling Europe, n.d.). These and other Japanese words and expressions are used throughout the rest of this chapter, instead of their translations. As in other areas concerning lean management, keeping the original Japanese terms conveys more accurately the concepts originated in the Toyota Production System (TPS) than do their translated versions.

10.2 Origins and current use

Firsthand observation has been part of lean management from its origin, especially as viewed by Taiichi Ohno, a pivotal figure for the TPS. An anecdote attributed to Mr Ohno (Wakamatsu, 2009) recounts that he asked a young engineer to draw a circle with chalk at a spot on the Toyota shop floor and to stand in the circle. The engineer stood for hours in the circle until the end of the day (Ohno scolded the engineer for briefly leaving the circle to go to the restroom). At the end of the day, Ohno asked the engineer: 'Have you figured it yet?' to which the baffled engineer answered: 'I have no idea.' Ohno made the engineer stand in the circle again the next day. At lunch, Ohno asked him again: 'Have you figured it yet?' The engineer responded: 'Yes, there's a problem.' Ohno was not indulgent with the engineer: 'You told me that you had continuously improved the shop floor but it has gotten worse because of your instructions! If you know what the real problem is now, go and fix it right away.'

The anecdote encapsulates the essence of going to *gemba*. The engineer had to see the shop floor with his own eyes and from close up. The problem was discovered and the answer was figured out by the engineer. The understanding of the problem and the proposed solution were part of the same thought process. In addition, the engineer learned many lessons, not only about the shop operation, but also about the self-directed essence of true problem solving – and Ohno's famed propensity for unadorned speech. Examples of *genchi genbutsu* go further than Ohno's practical lesson. A frequently cited example from Toyota (e.g. Padgett-Russin, 2003; The *Economist*, 2009) concerns the redesign of the 2004 Toyota Sienna. Yuji Yokoya, the engineer charged with the redesign, went to the *gemba*. He drove 53,000 miles across North America and figured insights that would be difficult to obtain if not by direct experience. Among other issues, he realised that Canadian roads have higher crowns in the mid-section than in Japan, probably to deal better with the large amounts of snow in Canada; that winds in Mississippi were very strong and demanded special attention to car stability; and that eating and drinking while driving is common in North America. Each discovery provided an opportunity for improvements in issues such as the minivan's drive handling, geometric proportions and 14 cup holders and a food mini-tray, among others. It is doubtful whether Mr. Yokoya would have been able to detect these issues by indirect information means alone. It is also arguable that he might have not appreciated their significance if someone had simply told him about them.

The usefulness of 'going to *gemba*' is manifested in many other circumstances. For example, New York Police Department Commissioner William J. Bratton used the concept to turn around the New York Transit Police in 1990. As reported by Abilla (2007), Bratton discovered that none of the senior staff officers rode the subway. To make officers aware of the safety concerns at the time, he required them to ride the subway to work, meetings and at night. The firsthand 'go to *gemba*' experience has been credited with helping to improve the approach to policing the subway and the substantial improvement in riders' safety that the city has enjoyed since the 1990s.

10.3 How to go to *gemba*

A *gemba* walk is not a stroll by the project, and not even a project walkthrough to assess its status. It has the basic objective of understanding reality through direct observation and checking information firsthand. Going to *gemba* requires an understanding of the operation being observed and the skill to meaningfully observe it. There is no single, cut-and-dried way to perform a *gemba* walk. Toyota Chairman Fujio Cho summarised the process of going to *gemba* most concisely by three principles: '(1) go see, (2) ask why and (3) show respect' (Shook, 2011). These principles have been interpreted and applied with different emphases, depending on the industry context. The following steps have been discerned from Mr. Cho's principles and successfully applied (Shook, 2011). Despite the subjectivity of a *gemba* walk, there are best practices that can be summarised as follows (Flinchbaugh, 2011):

- *Identify your purpose.* A *gemba* walk is performed to learn something about the situation at hand. If these two basic questions cannot be answered, the walk should not start in the first place: (1) Why are you observing the *gemba*? (2) What are you trying to learn from your walk?
- *Know your* gemba. The typical construction customer realises value at the project, and there is a natural tendency to equate *gemba* with project. But, the gemba of a particular situation can be elsewhere. Physical, direct observation is the key, for example, to understanding the interaction among subcontractors or the dynamics underlying a problematic material supplier.
- *Observe with a framework.* Now that you are at the *gemba*, are you capable of visualising the operations and their flow? Can you see the flow of the activities and interactions? Performing a gemba walk requires a baseline of technical understanding, management proficiency and ability to process the results of the gemba walk. In other words, there must be a framework to observe and not only look at the situation.
- *Validate what you see.* According to Wakamatsu (2009), Taiichi Ohno liked to share the story of a company's president who did not have a strong technical background and yet made a rule of visiting the company's shop floor to check the documents discussed in company meetings. He 'often discovered that what had been told in the [company] meetings were false assumptions'. Lean management relies on trust and reliability, which makes even more relevant the Russian adage that Ronald Reagan often used: trust, but verify (Watson, 2011).

10.4 Management by walking around

A management technique known as 'management by walking around' (MBWA) is sometimes confused with gemba walks. MBWA in its current form is attributed to the Hewlett Packard Corporation in the 1970s, and was popularised by T. Peters and R. Waterman in their 1982 book, *In Search of Excellence* (Peters & Waterman, 1982). The Business Dictionary summarises key aspects of MBWA as

> Unstructured approach to hands-on, direct participation by the managers in the work-related affairs of their subordinates, in contrast to rigid and distant management. In MBWA practice, managers spend a significant amount of their time making informal visits to work area

and listening to the employees. The purpose of this exercise is to collect qualitative information, listen to suggestions and complaints, and keep a finger on the pulse of the organization. It is also called management by wandering around.

(Business Dictionary, n.d.).

From this definition, it can be seen that a key difference between *gemba* walks and MBWA is their purpose. While both involve the personal presence of the decision maker, a *gemba* walk is more purposeful than MBWA. *Gemba* and *genchi genbutsu* gather information as part of a problem-solving strategy. MBWA attempts to find the problem and improve a manager's accessibility and company morale.

10.5 Case study

Higher education has had a long and fruitful cooperation with LC. Its initial developers were academics with outstanding experience in construction productivity improvement, and to this day important stakeholders such as the International Group for Lean Construction (IGLC) have substantial academic composition. These origins have resulted in giving education and training a central role. Hands-on, practical educational approaches such as simulations and roleplaying have been preferred methods for teaching LC, since these pedagogical tools effectively tie key concepts of LC together (e.g. Gonzalez *et al.*, 2014; Friblick *et al.*, 2007). The logistics of teaching become more difficult when a practical orientation is introduced into a classroom setting, compared to the well-established realm of lecturing and other traditional educational approaches. While LPS has been covered by traditional textbooks (e.g. Halpin & Senior, 2010; Forbes & Ahmed, 2010), research points to LPS being best learned by project teams currently applying it on a project (Gonzalez *et al.*, 2014). The old adage of 'learn by doing' applies to this case. The usefulness of going to *gemba* could be expected as a corollary.

The case study presented in this chapter introduces the experiences of introducing the LPS in an undergraduate construction management (CM) programme at California State University, Fresno (Fresno State). This case represents not only key lessons on implementing the LPS into undergraduate education, but it also acts as a proxy for the programmatic implementation of LPS on an actual construction project. The proxy is represented through a hands-on laboratory project to construct a simple wood-framed shed. Students were required to use LPS to manage the shed construction in this laboratory course, which provided a means to teach undergraduate students LPS in a manner similar to a professional setting. *Genchi genbutsu* led to an evolution of teaching strategies that could not be figured from the instructor's office.

10.5.1 Teaching the Last Planner System

There are important reasons to teach LPS in an undergraduate construction course in addition to its fundamental role in LC. The system can be used as a framework to teach many of the basic scheduling principles and practices expected of any construction graduate. It integrates master planning (a 'big picture' milestone scheduling), phase planning (collaborative schedule development), look-ahead planning (short-interval scheduling), weekly work planning (production management), and learning (project controlling). Each of these LPS steps maps directly to key learning outcomes commonly taught in construction scheduling and project management courses (Hyatt, 2011). LPS is also an excellent tool for teaching students the importance of continuous learning. LPS provides a systematic method of 'countermeasures' that encourages continuous learning throughout the life of the project. Thus, by applying LPS, students not only learn a system for continuous improvement, but they also have the opportunity to actually improve their performance on a class project using this approach.

There are important differences between teaching lean construction in a professional setting and in an undergraduate education setting. In a professional setting, it is assumed that participants have some level of experience and understanding of how a traditional construction project operates. That is not the case for many undergraduate students. Educators must assume that undergraduate students have minimal (if any) experience of a construction project. This complicates teaching a 'practice-focused' LC process in undergraduate education. For example, it is unrealistic to expect undergraduate students to create a realistic construction schedule for a large commercial construction project for the simple reason that they may never have worked on this type of project. How would they know the key steps in the construction process? How would they know all the terminology used in this type of construction? How would they be able to validate the accuracy of estimated durations for activities? Such an experience-based exercise would be very challenging for even the most intelligent student. Despite these inherent challenges, it is the instructor's opinion that LPS should be included in undergraduate construction higher-education programmes. LPS is becoming one of the primary tools used on LC projects in the United States (McGraw-Hill, 2013). Students must be prepared for a key tool that they will need when they graduate.

10.5.2 The construction project controls course

The construction project controls course at Fresno State incorporates three broad topics: project management, project accounting and construction business management. The course meets for two lecture hours and three laboratory hours each week. The lecture period included all three topics, while the laboratory covers the first two topics. The laboratory period requires teams of four to six students to build and manage a small, hands-on project throughout the semester. The laboratory is broken down into four phases: bid, pre-construction, construction and close-out. The primary focus of the laboratory is to have student teams manage the cost, schedule and quality of the project. The student teams are primarily graded on their team's ability to manage the project cost and complete the project on schedule to the identified quality standards. This focus adds ample complexity to the project since teams are required to complete most of the management work during the three-hour lab period.

The bid phase is one laboratory period in length and requires student teams to submit a bid (i.e. cost estimate) to complete the project. The project is 'procured' via a low-bid, stipulated-sum contracting process. The following three-week pre-construction phase requires students to create an initial plan for the project. Student teams create a series of detailed planning documents. The phase culminates with a pre-construction presentation that simulates a project team's presentation to their 'company executives' in which they demonstrate their collective understanding of the project. The initial approach to begin teaching LPS in the CM curriculum at Fresno State was typical in higher education. The instructor compiled content on the subject from a variety of trusted sources and then integrated this content into the course. This approach was made slightly more challenging in this case since most content was focused on professional or graduate education, or both. The instructor had to adjust the content to fit the specific educational outcomes for the undergraduate course, resulting in a very hands-on and ad hoc instructional approach.

LPS is introduced in the construction phase of the laboratory hours, which covers the majority of the semester (approximately ten periods depending on the semester). Student teams are required to build and manage the construction of the project during this phase. The primary purpose of this phase is to manage the project. A secondary purpose of the phase is to provide students with a further understanding of means and methods of the project. Each student rotates through three or four management roles. These roles include project manager, project engineer (optional), quality or safety manager and superintendent. The superintendent was the point person for most of the LPS activities in the laboratory. The role of LPS is discussed in detail in Section 10.5.3. The final phase is close-out, which requires the teams to submit all final payments and other close-out documents. This phase takes place during the final laboratory period of the semester. The primary purpose of this phase is to remind students that the project is not fully complete until all documents are finalised and accepted by the owner. The instructor also takes the final portion of the laboratory period to conduct a plus/delta exercise. This exercise has the structure of the typical closing activity in LC meetings, and allows students to provide feedback on what went well (plus) and what could be changed for the future (delta).

10.5.3 Conducting the laboratory

The laboratory project has been in place in its current form for the past three semesters. It requires student teams to build a small, wood-frame building. This scope was intended to make the project simple enough that student teams could complete the project in approximately 30 contact hours (ten weeks at three hours per week). It was further stipulated that the instructor could challenge students by adjusting project plans as desired throughout the semester. But, despite an apparently sound implementation plan, the laboratories experienced significant problems in the first two semesters, resulting in an incomplete and even frustrating experience for the students, the laboratory teaching assistant and the instructor. The first semester was a 'just do it' phase. Student teams were primarily focused on completing the project as quickly as possible and thus soon lost focus on the most important aspect of the laboratory, namely managing the fieldwork. The instructor added a mandatory requirement to complete the Weekly Work Plan (WWP) meeting in order to address this issue. This meeting forced the student teams to slow down and review the key production management aspects of the project. The student teams did complete the required meetings, but the documentation for each meeting was not consistent across the teams. The key benefits of LPS, namely collaborative planning and continuous learning, were not fully identified by the teams. At the end of the first semester no units had been completed.

In the second semester, the instructor attempted to address inconsistent LPS documentation and improve the quality of the WWP meetings by creating standardised templates for all elements of the system. The forms were created and shared on Google Drive to allow easy access for students. This approach introduced better standardisation for the WWP meetings

Figure 10.1 Image of work during the first semester

Figure 10.2 Image of work during the second semester

and a more consistent use of LPS on the project. Despite the availability of the forms, several teams failed to use all of the documents on time for each simulated week. Some teams would update the documents, but not use them during their WWP meetings. Other teams would refer to the documents, but not fully complete them each week. Ultimately it was an improvement from the first semester, but the online documents did not fully address the issue of consistency. At the end of the second semester, two of the four units had been partially completed.

10.5.4 Improving the laboratory experience by going to *gemba*

Improving the laboratory experience was a primary focus for the instructor between semesters. This seemed simple enough, but posed several key issues in order to implement LPS in a meaningful and effective manner. LPS is a practice-based lean construction process. Although it was created in an academic setting, the best practices have been developed by industry members. Many of these best practices are documented and shared by the Lean Construction Institute (LCI) at local, regional and national meetings. Therefore, the most current ideas and practices are improving on a continuous basis. This presents an interesting issue for an academic – how to ensure that students are presented with a relevant laboratory experience, if the process is continually improving. Even more important, LPS is created in such a way that it can be easily adjusted and utilised on any project. This simply means that the context of the project dictates the specific ways the LPS will be used on the project. Therefore, context is a critical aspect of using LPS in this laboratory setting. Context presents another issue for effective implementation of LPS in a laboratory.

Thus, the best way for the instructor to fully understand the current practices and effect of context was to go out to projects and see how the team members were implementing LPS. The instructor was able to better understand LPS by 'going and seeing' how it was utilised on several projects. Some observations from the *gemba* using a framework include:

- *Identify your purpose.* The purpose of *genchi genbutsu* was to identify the ways to better implement LPS in an educational laboratory setting.
- *Know your* gemba. In order to achieve this purpose, the instructor chose to visit four different construction projects (*gemba*) in order to observe how each project implemented LPS. These projects were of varying sizes and scopes, but were all building projects. Thus, the instructor was familiar with building projects from prior industry experience. This allowed the instructor to focus on the implementation of LPS and not necessarily on the technical aspects of the project.
- *Observe with a framework.* The observation framework utilised was a series of interviews with key team members (project manager, superintendent and foremen). Additionally, the instructor observed a series of meetings related to LPS.
- *Validate what you see.* Through the interviews the instructor noticed several key 'best practices' that were not typically identified in the basic LPS literature and education sessions. The best practices were then validated through observing the key meetings related to LPS.

A best practice identified by the instructor during *genchi genbutsu* was the lack of technology to implement LPS on the construction projects. Initial interviews with the team members found that using hard copies of forms and documents posted in prominent areas assisted the team in continuous and consistent use of LPS by the project team. Observation of LPS meetings validated this idea when there was almost no use of technology in most of the LPS meetings (especially at the field level).

10.5.5 Results of implementing visual control boards

This powerful best practice completely transformed the instructor's view of LPS implementation in the laboratory. In the initial semester, there was considerable reliance on various software systems (such as spreadsheets, scheduling software) to complete the LPS process. When the instructor reflected on the use of technology in the laboratory, it became very evident that students were focused more on learning the software platforms and less on using the system to improve performance in the laboratory. Thus a root cause of failures in the initial semesters could be traced back to the heavy reliance on technology to manage the process. During the third semester, the instructor printed hard copies of all documents and posted them on a 'visual control board' for each team. These boards were located adjacent to the laboratory site and easily reviewed by the teams at any time. It presented an ideal tool for student teams to use on a consistent basis when conducting their weekly planning meetings. Ultimately, the boards were consistently updated since every team member was accountable in a visible way. This method of documentation led to the successful and consistent implementation of LPS.

The improved coordination and productivity also lead to another key success – completion of the sheds. All sheds were fully completed by the student teams during this semester. Completion of the sheds not only provided a tangible result for students, but it also validated the importance of effective implementation of LPS in the lab. Finally, the overall learned outcome of effectively managing a small construction project was met through the completion of this project.

Figure 10.3 Image of the visual control board

Figure 10.4 Image of completed sheds of the third semester

10.6 Summary

The practice of *genchi genbutsu* provides the basis for effective problem solving on any project. It effectively allows the project team to identify the root cause of the problem in the specific context of the issue. This ensures that the team leaders do not try to present solutions that are generic in nature, and that may not solve the actual problem. The power of this approach comes in the leader's ability to view the issue at the site (gemba), identify the problem and work with the team to determine a solution that addresses the root cause. The case study presented in this chapter shows how the instructor utilised *genchi genbutsu* to solve the problem related to inconsistent use of LPS on a small construction project in a lab setting. The instructor decided to visit several construction projects that were effectively implementing LPS. The result was the identification of a key difference between these projects and the laboratory project, which was the minimal use of technology to implement LPS on the construction projects. The instructor adjusted the implementation in the laboratory setting and eliminated much of the software requirements for tracking LPS data and instead required student teams to use visual control boards. This small change in LPS implementation had profound effects, specifically that LPS usage increased dramatically and the construction projects (small sheds) were completed for the first time since the laboratory's inception. This simple example shows the power of 'go and see for yourself' on the overall problem-solving process. It is extremely difficult to determine a solution if you do not see the problem firsthand. Once you have gone to *gemba* and observed, it becomes much easier to determine the root cause of the issue. At this point, true problem solving can begin.

References

Abilla, P. (2007). Genchi genbutsu and tipping point leadership. Retrieved from www.shmula.com/genchi-genbutsu-and-tipping-point-leadership/430.

Ballard, H.G. (2000). The last planner system of production control. Doctoral dissertation, University of Birmingham, UK.

Business Dictionary (n.d.). Management by walking around (MBWA). Retrieved from: www.businessdictionary.com/definition/management-by-walking-around-MBWA.html.

Eisenhower Presidential Library (n.d.). Address at Bradley University, Peoria, Illinois, 26 September 1956. Retrieved from: www.eisenhower.archives.gov/all_about_ike/quotes.html.

Flinchbaugh, J. (2011). Going to the gemba. Industry Week Lean Learning Center Blog. Retrieved from: www.industryweek.com/companies-amp-executives/going-gemba.

Forbes, L.H. & Ahmed, S.M. (2010). Modern Construction: Lean Project Delivery and Integrated Practices, CRC Press, London.

Friblick, F., Akesson, A. & Leigard, A. (2007). Learning lean through lean game: A case from the infrastructure industry. In: C.L. Pasquire & P. Tzortzopoulos, *Proceedings of the 15th Annual Conference of the International Group for Lean Construction*, East Lansing, MI, 18–20 July, pp. 475–484.

Gonzalez, V., Senior, B., Ingle, J., Best, A., Orozco, F. & Alarcon, L. (2014). Simulating lean production principles in construction: A Last Planner-driven game. *Proceedings of the 22nd Annual Conference of the International Group for Lean Construction (IGLC-22)*, Oslo, Norway, 25–27 June, pp. 1221–1232.

Halpin, D. & Senior, B. (2010). *Construction Management*, 4th edn, John Wiley and Sons, New York.

Hyatt, B. (2011). A case study in integrating lean, green, BIM into an undergraduate Construction Management scheduling course. *Proceedings of the Associated Schools of Construction 47th International Annual Conference*, Omaha, Nebraska, 6–9 April.

Imai, M. (1997). *Gemba Kaizen: A Commonsense Low-cost Approach to Management*, McGraw-Hill, New York.

Liker, J.K. (2004). *The Toyota Way: 14 Management Principles from the World's Greatest Manufacturer*, McGraw-Hill, New York.

McGraw Hill (2013). *Smart Market Report*, McGraw Hill Construction, New York.

Padgett-Russin, N. (2003). Toyota takes to road to improve new Sienna. *Chicago Sun-Times*. Retrieved from: www.highbeam.com/doc/1P2-1479798.html.

Peters, T. & Waterman, R. (1982). *In Search of Excellence: Lessons from America's Best Running Companies*, Harper & Row, New York.

Shook, J. (2011). How to go to the gemba: Go see, ask why, show respect. Retrieved from: www.lean.org/shook/DisplayObject.cfm?o=1843.

The Economist (2009). Genchi genbutsu: More a frame of mind than a plan of action. 13 October. Retrieved from: www.economist.com/node/14299017.

Toyota Material Handling Europe (n.d.). The Toyota way. Retrieved from: www.toyota-forklifts.eu/en/company/Pages/The%20Toyota%20Way.aspx.

Wakamatsu, Y. (2009). *The Toyota Mindset: The Ten Commandments of Taiichi Ohno,* Enna Products Corporation, Bellingham, WA.

Watson, W. (2011). Trust, but verify: Reagan, Gorbachev, and the INF treaty. *The Hilltop Review,* Vol. (5)1, pp. 22–39.

Conclusions

Concepts and new learning frontiers

Value and waste forms the link between lean principles and tools that have been presented on this book. The realisation of the positive contributions of this book requires a change in the practice of construction management; and such a change demands 'seeing the whole' instead of the isolated parts of a system. This approach requires linking ideas holistically in order to explore the fuzzy borders between the concepts that have been presented in the book.

Imagine the delivery of a building project based on the traditional (non-lean) paradigm of construction management used for operationalising design and management actions. This project, Project A, involves the design and construction of a single-storey residential building. A representation of this delivery without lean construction would simply show the design and management actions as the inputs, followed by the outcomes (Exhibit 1). Given that construction is actions undertaken by contractors and their supply chains, the specialist skills and knowledge of the project actors are crucial to produce the desired buildings and infrastructure. Each construction project has a timeline that must be accomplished by contracting parties. The actions required from the contracting parties are diverse and involve design and management decisions within the project team, inclusive of the client. In short, construction management is ensuring that the design and management actions undertaken by the project team are done within the allowed time, effectively and efficiently. The delivery of Project A is therefore subject to the 'twists and turns' of the practice of construction management.

When the intricacies, which may include project complexity and uncertainty, overwhelm the project team and the adopted processes, underperformance would manifest, although the project could be delivered. Despite recent advances in the practice of construction management, complicated actions and actors imply that it is impossible to correctly forecast and realise performances in Project A. If the completed actions fail to deliver the project to the satisfaction of the client and the stakeholders, it may seem that construction management does not work or construction management has not been implemented properly. Maybe the design actions were not error-free, or maybe the management actions led to excessive rework on site. Although the delivery may include some measures of quality and good implementation, it can be difficult to determine what aspects (actions) should be improved to eliminate underperformance unless there is an efficiency-driven philosophy that supports the process.

Exhibit 1 Implementation of a construction project without lean

Such a philosophy is lean construction. A delivery using the lean approach could identify how construction management has worked and what intermediate outcomes need to be achieved for the actions to work better. This will bring the project actors to the point where implementation failure (not done right) and design failure (done right, but failed to work) are clearly differentiated. Exhibit 2 represents a project delivery pathway that articulates the causal mechanisms involved in producing two changes (changed action status – less waste and variability – and changed delivery status – enhanced performance). The first change relates to project actors' willingness to engender less waste and variability while managing people and machines in construction. The second change is the impact of their actions. For many construction projects, it can be helpful to articulate these changes so that notable underperformance can be avoided.

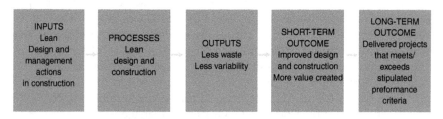

Exhibit 2 A simple pipeline and outcome waste elimination logic model

Understanding these two simple logic models in the context of a hypothetical Project A can be further enriched with cross-linking of concepts in this book. At this stage of the book, it is hoped that the reader who has followed the book in sequence will have a good overview of all the major concepts, allowing them to comprehend the nature of each. With this knowledge, the reader is equipped to explore the links between the concepts as illustrated in a concept map (Exhibit 3).

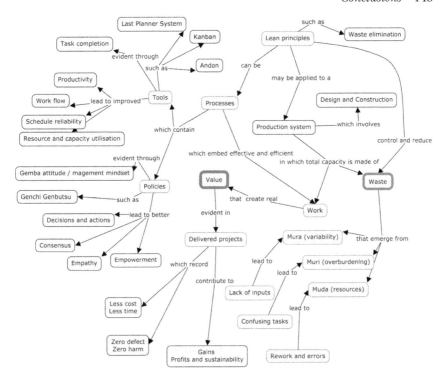

Exhibit 3 Linking concepts of value and waste in lean construction

Part I of this book has shown that lean construction can lead to less work and waste, which constitute the production system at a given time. Waste emerges from variability, excessive burden and resources owing to the lack of necessary inputs in the production system. The absence of such inputs could lead to chaos, making do, rework and workflow interruptions on a project. To avoid these production problems, lean construction as a philosophy is able to use a process in which tools and management policies are able to enhance created value.

As documented in Part II of this book, value can take objective, subjective and intersubjective form based on where it is situated. Nevertheless, value is always evident through performance criteria that have been set at the initiation of a project. Such parameters may be zero harm, which means the health and safety of both workers and the general public must not be compromised as a result of the delivery of a construction project. The value so created in a project will be influenced by the waste controlling tools and workplace (*gemba*) attitude used in the project.

Part III of this book highlights the utility of andon, LPS, and kanban in this regard. The links in Exhibit 3 show that these tools/techniques are able to reduce waste, albeit in varying degrees. For instance, LPS is able to promote the stabilization of flows, and in so doing it directly controls variability and indirectly controls resource utilization and overburden. *Genchi genbutsu* requires supervisors and managers to see the work, firsthand, and make better improvement decisions.

The importance of the abstraction in this concluding note is to foster learning regarding the waste elimination model of lean construction. The question is 'What are the new frontiers of learning for value and waste in lean construction?' As highlighted in several chapters of this book, the application of lean in construction requires pragmatic considerations. Purposeful elimination of waste in construction is not about formulaic applications. Rather, it is an iterative process that stimulates important questions which initiate answers that could improve theory and practice. It is hoped that readers of this book will be motivated to undertake research and reflect on practices in ways that will extend the frontiers of the waste elimination model of lean and advance approaches for the propagation of value in construction.

In order to make the transition from classical construction management to the lean construction paradigm which mandates waste elimination efforts, learning must occur in both higher education and industry. If the industry is to truly replicate the success of the Toyota Production System, it is important that novice and established professionals understand the fundamental concepts and principles that drive waste-elimination efforts and value-adding activities. In particular, significant opportunities exist to further improve the understanding and removal of waste through lean construction in terms of answering questions that are meaningful in practice. Some questions to be answered include:

- What are the multiple causal pathways for waste proliferation in design and construction?
- What is the influence of irrational client decisions on value and waste in construction?
- How do choice alternatives affect workflow and standardisation in construction?
- What is the influence of context on lean construction tools/method utilisation?
- Which measurement methods are suitable for tackling unobservable waste in construction?

Glossary

Numbers in parentheses following glossary entries refer to sections or whole chapters in which the concepts/terms are described.

Andon	The andon is the basis for 'jidoka', which means 'machines with human intelligence'. It is a management tool of visual control that shows the operation status in a workstation (9.2).
Buffer	This is an activity that reduces a shock or that forms a barrier between incompatible work processes. It is considered necessary waste to absorb incoming variation and thus an approach to reduce variation in the production workflow to increase on-site productivity (7.4).
Gemba	In Japanese, the word means 'the real place' or 'the actual place'. The gemba of a construction project from its owner's perspective are the points at which the project is actually built, all other places and steps being of importance to the contractor. Gemba is increasingly used as synonymous to genchi genbutsu, especially when used as 'going to the gemba' (10.1).
Genchi genbutsu	This is a Japanese term that refers to 'seeing for yourself'. It is one of the five principles of the 'Toyota Way', which enables the problem to be seen firsthand (1.4). In Japanese, genchi means 'actual place' and genbutsu means 'actual thing'. The combined phrase is frequently translated as 'going to the place to see the actual situation for understanding' (10.1).
Kanban	In Japanese, this word means 'card' or 'sign', and is the name given to the inventory control card used in a pull system (8.1). Kanban is a technique.
Last Planner System (LPS)	LPS is a production planning system designed to produce predictable workflow and rapid learning in programming, design, construction and commissioning of projects (7.1).
Lean construction	This is essentially the application of a new production management philosophy in construction. The

application sets clear objectives for the design and construction of projects with the overall aim of maximising performance to engender customer value.

Lean thinking
This is a business mechanism which is intended to provide a new way to think about how to organize human activities to deliver superior benefits to society and value to individuals while eliminating waste (4.1).

Look-ahead schedule
This schedule focuses on make-ready activities, thereby ensuring that the activities can be completed. By removing all obstacles, the look-ahead schedule ensures that it is possible to complete the activities in the determined sequence, according to the phase schedule (7.1).

Making do
This is a type of waste that occurs when a task has been started before all the preconditions for such an activity have been met. It occurs to keep capacity busy, although it usually has detrimental side-effects, such as an increase in work in process, a need for rework and creation of H&S hazards (1.3). It implies that work-as-done is different from work-as-imagined in plans (2.1).

Master schedule
This schedule is the main schedule which contains the deadlines and milestones for the entire construction process. The lower-level schedules focus on ensuring that the deadlines in the master schedule are adhered to (7.1).

Muda
This is a Japanese term for a waste of resources, in the form of a focus on non-value-adding activity and/or work that creates waste. Such activity is exposed through variation in output (1.3; 7.4).

Mura
This is a Japanese term for unevenness/non-uniformity, which is variation in work output within a production system (1.3; 7.3).

Muri
This is a Japanese term for overburdening, which is described as unreasonable demands on employees or processes, in the form of high rates of work or non-familiarity with work to be done (1.3; 7.5).

Negative variation
This term refers to finishing a work task after the deadline, which creates delays and disruptions to the schedule (7.2). It is undesirable in a production system.

Percentage Planned Completed (PPC)
This is a measurement tool. The measurement is a simple comparison between the planned completed and the actual completed. In this way the activities which were not completed according to the schedule are identified. The tool provides feedback to the site

	manager, who can easily determine whether he needs to intervene.
Phase schedule	This schedule focuses on identifying in what order the activities should be completed. At the outset the deadlines and milestones are defined in the master schedule. The sequence is determined by working backwards from the deadlines and identifying activities and handoffs between the work crews (7.1).
Positive variation	The term refers to finishing a work task before the deadline, and it creates gaps in the production and results in unexploited capacity (7.2). It is undesirable in a production system.
Production system	This is the sum of work and waste in a system (1.2).
Resilience	This is defined as the intrinsic ability of a system to adjust its functioning prior to, during or following changes and disturbances, so that it can sustain required operations even in the presence of continuous stress (2.3).
Resilience engineering	Developing key tools and methods for the maintenance and management of safety. This is a safety management paradigm that explicitly values the positive side of variability (2.1).
Stabilised workflow	This flow occurs when resources and capacity are matched and decreased variation (*mura*) and increased productivity are observed in production. A stabilised workflow is achieved by reducing inflow variation and by keeping a backlog of ready work to absorb incoming variation (7.3).
Value	This term is seen as a relation established between subject and object. It may form a referential relation. Lean construction understands that value to the client is the variable that defines the visibility and quality of certain product attributes (4.3). For lean construction, the concept of value focuses on matching all customer requirements in the best way possible (design and production) (5.4). This is the price a client pays for a required product or service (1.2). It could stand for a person's willingness to pay the price of a good in terms of a cash return for certain product benefits, meaningful difference and action (6.2). Value is understood as the fulfilment of the demands and requirements stated by the end customer (7.1)
Value-adding activities	These are activities that contribute to the progress of work and for which clients are prepared to pay (6.2).
Value-supporting activity	This is a work that does not add value, but is necessary under an operating condition (1.2).
Variability	This is the quality of non-uniformity of a class of entities, which can be designed into a system. It is

also the range of performance measurements, values or outcomes around the average which represents all the possible results of a given process, function or operation (2.1).

Variation

This term is associated with variability and it is observed when the production throughput results in activities being completed either before or after the scheduled deadline (7.2).

Waste

This is an activity that consumes time and resources but fails to add value to the final product. Such work does not add value and is not necessary. It may also be an unwanted physical functionality of the product, as well as the use of more resources than is needed, or an unwanted output (1.2).

Weekly work plans

These plans contain the work which has to be carried out on the construction site during the following week. The weekly work plans detail the work tasks that should be completed within the work week and by when they must be done. When selecting work tasks, only work tasks from the backlog of ready work are selected. This ensures that all selected work tasks can be carried out (7.1).

Index

Entries in *italics* refer to titles of documents.